CAMERON'S ENGINES

More Stories from the Support Crew

Michael Mather (editor)

AMBERLEY

Dedicated to the drivers, firemen and traction inspectors, many no longer with us, who we worked with over the years.

Also, to the thousands of passengers who chose to travel with us, and our fans at the lineside.

All profits and royalties from the sale of this book will be donated to Scottish railway preservation projects.

Front Cover: In pouring rain, No. 61994 *The Great Marquess* and No. 60009 *Union of South Africa* arrive at Wymondham on 24 June 2014, ready to take part in the following weekend's Mid-Norfolk Railway Gala. (Ian McDonald)

Back Cover: No. 61994 *The Great Marquess* and No. 60009 *Union of South Africa* double-head the inaugural 'Great Britain' rail tour on 10 April 2007, at Dalnacardoch on the climb to Druimuachdar Summit. (John Whitehouse)

First published 2025

Amberley Publishing
The Hill, Stroud
Gloucestershire, GL5 4EP

www.amberley-books.com

British Library Cataloguing in Publication Data.
A catalogue record for this book is available from the British Library.

ISBN 978 1 3981 2260 4 (print)
ISBN 978 1 3981 2261 1 (ebook)

Typesetting by SJmagic DESIGN SERVICES, India.
Printed in Great Britain.

Foreword
by Dr John Cameron CBE

So, No. 60009 has received a stablemate!

No. 61994 K4 Class *The Great Marquess* has joined *Union of South Africa* at their new home on Balbuthie Farm. Both now rest side by side in the new 'loco shed', which in due course will be open to the public.

The appearance of No. 61994 on the scene at Balbuthie was not as a result of a long-term venture come to fruition. In actual fact, it began with a phone call between myself and the present Lord Lindsay.

I first met the present Lord Lindsay when he was a minister responsible for agriculture at the Scottish office in Edinburgh. He telephoned me one day enquiring if I knew of anyone who might be interested in buying No. 61994. After a lot of thought, I spoke with him and said, 'I might know of someone' – and the rest is history!

I guess the deciding factor in my thinking was that with No. 61994 and Number 9 at my disposal there really was nowhere in the country I could not take a steam special, with *The Great Marquess* hauling over restricted tracks and *Union of South Africa* on the main lines. So it was for many years that the two locos were to be seen in almost every part of the country.

I'm often asked which of the two locos I enjoy driving most. I am going to 'duck answering that one', as the politicians say, because I got the same thrill and satisfaction out of both.

It required just as much skill taking *The Marquess* up Loch Treigside with a heavy fish train as it did to maintain 60 mph with the *Union* with twelve on ascending Grantshouse on the ECML.

Incidentally, none of my support crew will answer that question either, so the reader must draw their own conclusion.

In terms of highlights with the two locos, obviously hauling the royal train to Tweedbank with No. 60009 and Elizabeth II on board to open the Borders Line was a great occasion. However, the opening of the Alloa Line with No. 61994, which I have known for many years, gave me the same feeling of accomplishing a worthwhile task.

In summary, both these locos have been a big part of my life and it has been my privilege to own them and to meet with so many fine railway people. Their future is now secure as I have established the Cameron Railway Trust. Please come along and visit when the Railway Museum at Balbuthie is open.

Thank you all, particularly my support crews, for your lifelong support.

Introduction

Following the success of our last book, *60009 Union of South Africa: Stories from the Support Crew*, and the fact that both of Mr Cameron's engines are now safely tucked up at Balbuthie, we felt the time was right for another volume, this time including *The Great Marquess* as well as Number 9 within its pages. The previous book was so well received that we have been able to carry out our pledge to donate all our royalties and profits from sales to Scottish railway preservation projects, and to date, six projects have benefitted from our donations. We plan to do the same with this new volume and hope it does as well.

As before, we present stories portraying what it is like being on a main line support crew, the ups and downs and the fun we all had, even though it could be hard work at times. We have been raiding our photograph collections to present a well-illustrated volume and as before, we have included a selection of photographs from our friend and colleague the late Davy Murray and others that have been gifted to us over the years. We are also pleased to share the back story with you about Number 9's flight down Wellington Bank from the professional railwayman's point of view.

Although we are now retired from support-crew work, apart from two of our younger members who now work on No. 60103 *Flying Scotsman* and No. 6233 *Duchess of Sutherland* and some of our English members who volunteer on various other locomotives and heritage railways, we remain together as a group, still meeting regularly. Sadly, we are now one short, though, following Colin's passing in early 2023.

The Great Marquess was a fantastic addition to the fleet, being able to go to places that Number 9 couldn't, such as the West Highland Line, Kyle of Lochalsh and Blaenau Ffestiniog. It was also a popular locomotive on heritage railways, especially on steeply graded ones where its hill-climbing ability could be used to the full.

We were blessed to have these two fine locomotives to look after, sometimes both in one day, and it has all made wonderful memories for us, and it is great to be able to share these memories with you, the reader. Both locomotives are outstanding in their own ways and Number 9 for all the records it holds even managed another one on its last trip. It became the only steam locomotive to have crossed all three Forth Bridges at Queensferry when it crossed the new Queensferry Crossing on its way to Balbuthie, a record that is very unlikely to be broken.

We hope you enjoy the stories and memories we have shared with you here as much as we both enjoyed making them and writing about them.

Michael Mather

The Many Identities of Number Nine

On 23 March 1991, while carrying the Osprey nameplates, Number 9 worked an SRPS rail tour from Perth to Carlisle via the GSWR as part of a movement to Crewe for working North Wales Coast Expresses the following month. Nicknamed 'The Trans Gorbals Express' by the support crew, as it passed through Glasgow, it is seen here at Dumfries station along with the usual welcoming committee. (Peter Drummond)

In August and September 1991, while on North Wales Coast Express duty, Number 9 ran as long-scrapped Haymarket shed mate No. 60027 *Merlin*, much to the surprise of photographers at the lineside who were expecting *Union of South Africa*. Here, *Merlin* stands at Holyhead station about to leave for Crewe and tackle the 1 in 75 climb away from the port. (Davy Murray)

On 16 February 1992 Number 9 took part in the 150th anniversary celebrations of the opening of the Edinburgh and Glasgow Railway. The guest of honour was the Right Honourable William Whitelaw MP, the grandson of William Whitelaw, chairman of both the North British Railway and the LNER and after whom A4 No. 60004 was named. Number 9 is seen here posing as its long-lost classmate at Falkirk. (Jules Hathaway)

Once again carrying its real name, 60009 *Union of South Africa* accelerates away from Ladybank on 13 April 2008 at the head of The North Britain rail tour. It is heading for Dundee, from where the train was hauled by a Black 5 running tender first to Perth with Number 9 on the end of the train. Both locomotives then took the train to Inverness. (Michael Mather)

1993 Crew Training

In 1993 Number 9 was hired by ScotRail to train staff for steam running, the plan being for three steam inspectors, two drivers and twelve firemen to be trained. The instructor was Aberdeen-based Inspector Lachie Duncan, a vastly experienced steam man who knew Number 9 very well. These training runs ran Monday to Saturday and were open to the public with a flat fare of £15 for the circular trip from Perth through Stirling and Kirkintilloch to the eastern edge of Glasgow, turning left at Springburn and heading back via Cumbernauld to Stirling and Perth.

For the loco and support coach we started at Markinch and ran to Perth to turn on the turntable in the yard at Muirton, then pick up the six ScotRail-liveried Mark II coaches in Perth station. On arrival back at Perth, we would leave the coaches in the station then go to turn again and head for Markinch via the triangle at Thornton to turn again then go on to the shed.

The training should have started in the June 1993, but an incident occurred on the very first day during the engine and coach move from Markinch to Perth when the

Traction inspectors and trainees line up with Number 9 at Perth station. From left: Geordie Steele, Peter Annandale, Chris McGuire, Lachie Duncan, Robert Plant, George Beattie, David Bowie (music-loving readers will notice a rock-music theme going on here!) and Iain Royal. (George Beattie Collection)

steam pipe feeding the fireman's side clack valve failed sending steam into the cab. As a result, an inquiry was held, and the training was postponed. Four months later it started in the October and ran for four weeks. For me it coincided with a three-week period when I was sacked pending an appeal from my job (I was reinstated) and my wife told me in no uncertain terms to 'get out of her sight and play with the train for however long'! Who needs to be told twice?

The days were long and started with helping Fraser, one of the trainee drivers, to get the loco ready. He would sort the fire and I would do the oiling round. Most of the prep work was carried out the night before and I would top up the syphons and oil boxes. I can't remember if the trainees joined us at Perth or Markinch, I suspect it was more likely Perth but some at Markinch. Lachie joined us at Markinch to take the loco to Perth with Fraser firing as he was already passed out to fire. Smartly dressed Marilyn would always turn up to see us off and lock the shed door once the 'racehorse' had bolted.

The train was booked to stop at Stirling on the way out and Cumbernauld, where we took water. We stopped again at Stirling on the return and took water again there. The trains carried more passengers on the Saturday for obvious reasons and were quite lightly loaded during the week.

I spent a lot of time on the footplate watching and learning and one thing about Lachie's route knowledge struck me and stayed with me ever since. He always sat in the fireman's seat and spent most of the time looking in the cab watching the trainee driver and fireman. There was a rough piece of track in approximately the Dunblane area which always caused the loco to lurch quite hard. I noticed that Lachie, without looking outside the cab, would always stand up at one point and then the loco would jolt with a bang and he would sit down again. He knew exactly where he was without having to look outside the cab, now that is route knowledge.

The climb from Stirling through Dunblane up to the summit at Kinbuck is a hard one and on Saturdays John Cameron would do the trip and he always drove from Stirling to Perth. One day it was raining, and the rails were slippery, and Number 9 was slipping a fair bit on the climb to Kinbuck and Fraser was driving. Normally, when a loco slips the regulator is closed until the slips stops and then is reopened, but what Fraser was doing was leaving the regulator where it was and controlling the slip using the cut off. When the loco picked up its feet Fraser would shorten the cut off until the slip stopped then slowly let it out. I have never seen any other driver control a slip in that manner.

When we got back to Markinch shed in the evening, the rest of the support crew would be waiting and the work of disposing and prepping the loco would start. John Cameron brought along an old rickety tractor with a front bucket so we could load the coal. The lingering memory of it was the unfamiliar controls as the lever for the bucket was push down to raise it and push up to lower it! We got the hang of it quite quickly.

One week was lost as a repair to the fireman's side clack valve had to be made, which involved blowing down the boiler to lower the water level and removing the valve and steam key casting. Once the repair was made and the valve refitted the boiler was refilled and steam raised again.

To make up for the lost week we had another short stint in the January of the following year. Again, the trips were open to the public, but the coaches were hired from the SRPS

In the cab of Number 9. From left: trainee driver Iain Royal, trainee steam traction inspectors/firemen Robert Plant and George Beattie and instructor Lachie Duncan. The second trainee driver was Fraser Birrell, who along with George Beattie is also a member of the support crew. (George Beattie Collection)

at Bo'ness. That meant they were Mark I coaches in maroon livery. The week we ran was bitterly cold and frosty but very sunny. The stock was heated by the loco's steam heat, but the toilets were out of use with no water to prevent frost damage, so any hardy soul that ventured on the trains used our loo in the support coach. Anyone who took lineside photos of those trains would have captured the closest reproduction of the famous Three Hour Expresses between Glasgow and Aberdeen as the loco was filthy and hadn't been cleaned for months and it was hauling a rake of maroon Mark I coaches.

Time travel is possible.

Jules Hathaway

That Trip

Introduction by Jules Hathaway

A voice in the coach corridor attracted my attention, 'Good grief you do look like your brother.' I turned and saw a man I had never met before and looked at him with a puzzled expression. A hand came my way and the unknown man said, 'Hi, I'm Ieuan Davies. I know your brother Danny.'

Number 9 was on a tour from Clapham Junction to Exeter on the former LSWR route via Salisbury and Yeovil. The next day we were to go from Exeter to Paddington on the former GWR route via Bristol and Swindon and THAT trip on Sunday 19 February 1995 was to become the stuff of legends (see our first book, *60009 Union of South Africa: Stories from the Support Crew*).

Ieuan was the Operational Responsible Officer for both trips and he has written his story of how he became involved with steam footplate crews, giving an insight into the planning and what goes on behind the scenes that very few people including us knew about.

If he ever offers to make you a cup of tea, I hope you like it strong!

Over to you Ieuan...

In 1994 things were a bit strange in the railway industry. Things were gearing up for privatisation and I had recently joined the Operations Management team at InterCity GW. It wasn't very long after getting our new team together that we realised that we had a hole in our organisation as regards special trains and steam workings...all because some enterprising individual had put forward a bid to provide train crews for such workings. Normally in such circumstances the responsibility for this would have fallen on the Chief Traction Inspector, who reported to me. Whilst a thoroughly good chap, he had one glaring fault...he didn't like steam locomotives and knew that he couldn't be made to work on them! I remember him saying to me, 'You have to volunteer to work on steam...and I ain't volunteering!' We let things sort of swing with the local traction inspectors keeping an eye on things, but two events brought things into sharp focus, and it was clear that a tighter managerial grip was needed.

Firstly, Rail Express Systems, who were coordinating a lot of steam work on their Safety Case, needed more loco crews in the south of the country, which we were happy to provide, but obviously needed to manage from a competency point of view.

Secondly, there was an incident with No. 6024 leaving Bristol Temple Meads with an over-full boiler, which resulted in damage to the locomotive and a substantial bill to InterCity GW! Following this, my boss came to see me, actually she took me out to lunch, and passed the buck to me! I remember her comment, 'You like, know, understand and play with the bloody dirty things (steam locos), go get formally signed off on them and get a grip of what is going on...but don't you dare enjoy yourself!'

You get the picture?

Working with the rump of the InterCity Special Trains Unit at Euston I got to work with and gain the trust of our steam crews by being there alongside them, whilst also offering a robust communications channel with the loco owners.

All was good for a while and then the infamous *Blue Peter* incident occurred at Durham...and things had to change pretty quickly.

I was summoned to Euston by Steve McColl, the then Director of the Special Trains Unit. Now I know that the late Steve is a bit of a 'Marmite' character, but I have to say that, at the time, I found him to be very driven to deliver the best for both the industry and loco operators; let's leave it at that. The purpose of the 'summit' at Euston was to go through the findings of the investigation into 532's catastrophic wheel slip and to draw up measures to make sure there was no repetition of this. One of the outcomes was to ensure that loco crews were fully conversant and competent with the type/class of locomotive that they were working; it was no longer acceptable to view all steam locos as the same…thank goodness!

Looking at my notebooks from that period, I recall the following locos that my crews and I worked on in the couple of months after the *Blue Peter* slip: Nos 6024, 828, 60009, 46229, 70000, 80079 and 46521.

The only person on GW with experience on all of these classes prior to that was me! I became a very busy chap, arranging familiarisation handling and briefing on various heritage railways as well as getting crews onto light engine movements on the main line and meeting owners in different places. I was lucky to have a network of good contacts and mates, such as the late and very dear Ray Towell at the NRM, Alun Rees on the SVR, Steve Underhill of No. 6024 and a number of others who all 'got it' and were more than willing to help. My own background of a varied operations career, as well as working on heritage railways and support crews, helped massively.

Towards the end of 1994 we got the list for steam workings for early 1995 and Number 9 featured in it.

So, on 16 December 1994 I arrived at Didcot at silly o'clock in the morning with a bunch of Exeter and Bristol footplatemen to move Number 9 to Newport (via Gloucester) and to familiarise them with the delights of Sir Nigel's finest. I remember that there had been a problem with one of the loco's gradient pins and as a result we were very late off shed. By the time we arrived in Newport, I was reasonable happy that we had started to get the message over that this class of loco was neither a King nor a Bullied!

Let me provide a bit of background on the steam-crew rostering.

When we, ICGW, were called on to provide crews for special trains, be they steam or diesel, we always had to rely on either the goodwill of drivers to move shifts or work rest days. Obviously, the depots existed to crew our core timetabled services and really were not over resourced to enjoy much special-train activity. Fortunately, we had the wholehearted support of both the drivers and the regional ASLEF officers, indeed the depots involved always seemed to take a pride in supplying men and collectively watching their performances. Our depots at Paddington, Bristol, Exeter, Plymouth and Swansea were all involved, and we also saw other crews from RES working alongside ICGW men. It did mean, however, that when our Ops Resources staff at Swindon accepted a train to crew, our local managers and supervisors had to swing into creative action to cover the jobs. Fortunately, the head of the resources team reported to our department and actually had a desk only a few yards down at our office in Swindon. Plotting and scheming was our business!

When we started planning the various special jobs in early 1995, I put a marker on the 18/19 February quite early on, and, as I have previously said, we had started the familiarisation process with the men. My local Driver Standards Manager at Exeter, Graham Petrie, was duly tasked with covering the various crew diagrams and liaised closely with me once he had drawn up a list of those suitable. We, ICGW, had a good number of steam men at Exeter, and there were two 'stand-out' drivers, Gerry Gayton and Norman Helliwell. In addition, there was Trevor Barnett, a Regional Railways man whom we had used on a number of trips with a reciprocal cover arrangement. I knew Trevor reasonably well from earlier in my career when he was at Marylebone and we were working specials from there to High Wycombe and Stratford; I had actually fired to him on a Black Five on one Christmas Special working! When we looked at available drivers for the Number 9 job over two days, we realised that we could be a bit stretched. Saturday would, in my mind, be a precursor to the main event on the Sunday. For that day I had initially thought of Norman to drive and Dave Walker as fireman, both being ex-Leeds and York men. Unfortunately, Norman was on leave, though Dave was fine for the job on the Sunday. Although Gerry was an excellent loco man, he had absolutely no experience on an A4, nor for that matter any wide-box engine and by his own admission was a GWR man through and through! So that left Trevor, who was only too keen to be involved, particularly as he was soon to retire.

Saturday saw Brian Hayton driving and Dave Brown firing from Salisbury to Exeter. I remember meeting the train at Salisbury that Saturday lunchtime and then enjoying the trip down to Exeter via Yeovil. My notes don't really say much except that it was 'adequate'. The West of England Line at that time had numerous speed restrictions and funny operational instructions; I seem to recall that there was quite a severe and long TSR at Axminster due to damaged caused by flooding.

Here I am firing Number 9 a few weeks later on 9 March 1995 between Saltley and Derby. On the regulator is Saltley driver the late Maurice Norman and the blue cup to the right is being held by rostered fireman Ron Harper. (Ieuan Davies Collection)

On the next day Trevor, Dave, Graham and I met up at Exeter and discussed what lay ahead. I think I said something like, 'I want to see how bloody good an A4 could be!' We agreed that, with the owner's approval, we would close off the footplate to general visitors in good time to allow Dave to get the fire and boiler just how he wanted. I remember shunting the loco and support coach out of the bay siding and onto the train in Platform 6 at St David's and then that's exactly what we did.

I briefly saw Steve McColl on the platform, him having travelled down on the train, and I think that he might have had an inkling that something was about to happen. He told me that he was going to have lunch and would see me at Taunton.

Trevor got himself comfy in the driver's seat, grinning at the three of us and saying something like, 'You fellas better have your shovelling heads on!'

We got the right away and I picked up the shovel, putting a few around the box. I looked up and we were already stomping through Cowley Bridge Junction at just over 40 mph; I remember thinking, 'Bloody hell, that's a bit quick!'

Dave, Graham and I then just slipped into a nice rhythm, six or eight around the box, hand the shovel over. Trevor got on with driving, J. C. sat in the fireman's seat and various support-crew guys popped in and out of the corridor tender.

My notes said that we went into Whiteball Tunnel at 72 mph, which with twelve on for 413/430 tons was remarkable; a Class 47 couldn't do that with that load! The experience inside the tunnel was amazing. It was like the smokebox contained a load of angle grinders as sparks flew out of the blast pipes and streaked past the cab.

As we burst into daylight I glanced at the speedo, 'Hmmm, not bad.' If you ever get the chance to listen to Driver Duddington describing his run down Stoke Bank on *Mallard*, do so, that's how I felt. J. C. gave Trevor a bit of advice to get the best out of 9 and she fair flew along, just as Sir Nigel intended!

I recall Trevor shouting, 'Shall I shut off?' Both Graham and I were shaking our heads… until we felt that we had done enough. I think Graham asked J. C. if he liked what we were doing! J. C. grinned at me and yelled something like, 'Aye, but I like that more!' He pointed at the speedo.

I gave Trevor the nod and he whipped the regulator shut and got Number 9 into drift, whilst simultaneously starting to apply the brakes…Taunton arrived very quickly!

When we arrived at Taunton the scenes were incredible, unbelievable and so very, very special. I looked back from the tender cut out as Trevor brought 9 to a stand and the doors flew open all down the train. Once the initial clamour died down, I stood next to the tender and spotted Steve McColl…he gestured to me to go over to him and as he did so also slapped the back of his hand! Oooops! 'What the hell d'ya think you are doing?' he said to me, continuing, 'I was trying to enjoy my soup!' 'Don't you ever, ever pull a stunt like that again… unless you get me on the footplate with you!' At that he grabbed my hand and shook it firmly. 'Congratulations, you are the fastest steam men since Bill Hoole! Bastards!'

It was all a bit overwhelming at Taunton, but I knew that we had just done something a bit special and crazy. I was also worried that the loco might had been affected, but the support crew assured me all was well with her. I kept thinking about *Mallard*'s middle big end! It was a relief to get back underway and concentrate on the job again.

If the run down through Wellington was special, look at the performance between Taunton and Bristol. Consulting my log from the time, it was mile after mile of sustained

The 'fireworks' at Whiteball over for the day, in fading light, Number 9 approaches Worle Junction. (Nigel Hayman)

effort at almost constant speed on a gently climbing railway. All I can say is that it was more than impressive; stunning is the word that springs to mind.

Then things all went very pear-shaped in Bristol. First a broken hydrant, then a ballast train derailment on Horfield Bank, single line working to Bristol Parkway, greasy drizzle and slipping to a stand, hand-sanding the rails (yes, that was me) ... sixty-three minutes to cover 6 miles!

There was some further impressive running over the Badminton Line. Then the signal stop in the middle road at Swindon after Wootton Bassett HAB had been activated...I recall my telephone conversation with Swindon PSB. 'Hello, driver, you have a hot axle box activation on the third axle.' 'We are a steam locomotive!' 'Oh, and the thirteenth axle too.' No fault found...! A further twenty-two minutes lost.

Yes, it was a shame to have to take Number 9 off at Didcot, but operationally it made sense and then we ended up in the pub.

Don't even guess what I was like the next morning at our weekly Operations Performance Group Meeting!

You asked if there was any fallout or recriminations afterwards?

Well, not really, but there was an awful lot of jealousy from certain quarters. One Railtrack Signalling Manager tried to stir it up by claiming that he had a printout from the Wellington HAB detector showing a high speed as we passed, but as such devices were not calibrated as speed detectors it was pretty much thrown out! My boss at Swindon did, however, feel duty bound to send the (my) Chief Traction Inspector to ask Graham Petrie a few specific questions when I wasn't about. Amongst these was, 'Was the top speed that you attained 90 mph?' Graham answered honestly, 'No.'

Case closed.

Ieuan Davies

A Tale of Two Springs

For various reasons, broken springs, usually on the tender, are a bit of an occupational hazard, and of course they happen at the most inconvenient times.

Two such incidents are worth recording here. On Thursday afternoon on 23 April 1992 Jules and I set off by train from Fife to Wolverhampton, from where we would take a taxi to the Severn Valley Railway at Bridgnorth. The train trip was uneventful, but the following taxi ride was a real white-knuckle affair, the driver seemingly wanting to break some sort of record, but by the hot smell coming from the taxi, it obviously wasn't used to this sort of treatment. We didn't mind the speed all that much as it meant we would get to the Railwayman's Arms on Bridgnorth station (where the rest of the support crew would be waiting on us) before closing time. History doesn't record whether the taxi made it back to Wolverhampton!

A stop was made at Newport while on route to Hereford. The Scottish Steam Specials headboard was a remnant from rail tours Number 9 hauled in the 1970s. (Jules Hathaway)

The following day Number 9 and support coach were due to travel to Hereford via Worcester, Gloucester, Newport and Abergavenny. This long way around was due to the limited clearance through Colwall tunnel on the direct Worcester to Hereford line, a route that would have been much shorter. On the Saturday we were due to take over a north-bound Welsh Marches Express from West Country Pacific No. 34027 *Taw Valley*, which Number 9 would haul as far as Crewe. The trip to Hereford was the usual enjoyable affair that all engine and coach movements were.

On arrival at the Bulmer's Railway Centre where Number 9 was to be stabled for the night, the fire was attended to and left to die down a bit, the boiler filled up and a hose put into the tender which would slowly fill it up, little did we know what this was going to cause! We would leave cleaning and oiling until tomorrow as there would be plenty of time, or so we thought.

It was now time to get cleaned up and 'hit the town' for our evening meal. Just as we were setting off there was a terrible bang from the tender. 'What the **** was that?' said Davy, our RO for the weekend, as Lindsay was away in New Zealand. On looking round the tender, a broken top leaf on the driver's side leading axle spring was found. The increase in weight as the tender filled must have caused what must have been a weak leaf to fail. It looked like we might not be going anywhere tomorrow as we didn't have a spare spring with us.

This was in the days before we all had mobile phones, but luckily a volunteer at the centre whose house was next door allowed us to use his phone to contact the SVR to see

The broken spring that made the loud bang! (Michael Mather)

what could be done. With no spare there either, we found out that the only one available was *Flying Scotsman*'s spare, which was at Southall in West London, and that we could borrow it. The SVR general manager Alun Reece arranged to drive down overnight to pick up the spring, with the hope to be with us mid-morning, which he managed to do. With nothing more we could do that night we went for our meal. Saturday dawned and we set about removing the broken spring and preparing the engine for the trip ahead, Jules and I electing to clean the engine and coach, quite a task for two, our arms about hanging off by the time we'd finished. All was ready for the arrival of Alun with the spring, which was soon fitted.

Needless to say, word had got out about our predicament and the 'jungle drums' had been beating overnight, so those on the tour were very pleased to see us when they arrived at Hereford. The volunteer who had helped us, along with his wife, got a free trip in the support coach as a way of saying thanks, much to the annoyance of a certain BR official who told us we had too many on board, little knowing that without the help we got, there would have been NO Number 9 to pull his train.

Moving on to 2002, Number 9 was fresh from its second overhaul at the Severn Valley Railway, during which time it was fitted with air brakes and had one trip under its belt, to Plymouth during which there were a number of teething troubles with the air brake system, but these had now been cured.

With Severn Valley Railway General Manager Alun Reece at the coach door and Colin's reflection in the coach window, Marilyn, Davy and Nick chat to railway photographer and author the late Jeff Cogan on the platform at Birmingham International station, unaware at that point that Number 9 had a broken spring. (Michael Mather)

My view from the cab of the Class 66 as we arrived at Kidderminster. (Michael Mather)

Number 9 was now due to haul The Liver Bird rail tour from Birmingham International to Liverpool and return on 2 March.

Before leaving Kidderminster, the tender was topped up with coal and Nick and I were all ready to go up and trim the coal as it was well above the loading gauge, but those in charge thought otherwise, and so off we went.

No sooner had we got 'under the wires', there was an almighty bang and lumps of coal came flying past the support coach. The out of gauge coal had caused a flashover and so, at the first opportunity, we were put into a siding to trim the coal, which should have been done at Kidderminster, but who were we to argue?

Shortly before departure time at Birmingham International, we discovered a broken leaf on one of the tender springs, not a top leaf, fortunately. The decision was made to carry on, and as we didn't have a spare with us, a quick phone call back to Bridgnorth, we all had mobiles by this time, had engineers and a spring winging their way to meet us at Edge Hill depot, Liverpool, where Number 9 was to be serviced for the return trip – 'the organisation all goes like a well-oiled sewing machine' as Captain Mainwaring would say!

Liverpool was reached with no problems; the spring was changed and Number 9 serviced ready for the return trip.

Number 9 and its train were hauled back down to Lime Street station by a Class 47, which was to remain on the train for the return trip but would be dead. This meant that Number 9 would be faced with the 1 in 93 climb from a dead start up to Edge Hill, with eleven coaches plus the dead Class 47, the equivalent of a fourteen- coach train.

This didn't deter Crewe driver Ray Hatton, who was on his last steam turn before retirement. With just a single slip, he got Number 9 going and with great skill with the regulator and reverser made a superb climb up to Edge Hill. The three-cylinder beat echoing off the deep cutting sides, we had never heard Number 9 making so much noise. It was fantastic to hear, and we were quickly on our mobile phones to let support-crew members not present hear the three-cylinder music.

On arrival at Birmingham International we moved to Bescot depot to spend the night before returning to Kidderminster, the return trip to be with a diesel pilot.

The following afternoon an EWS Class 66 locomotive coupled onto Number 9 ready for the trip back to Kidderminster and 'The Valley'. I was lucky enough to be invited into the cab of the 66 for the trip, cause for a quick call to an at-home Jules to make him green with envy.

The rest of the year was to be a good one with Number 9 climbing the Lickey Incline and Shap for the first time and also having a spell working the Orient Express British Pullman.

Michael Mather

Davy Murray

A selection of photos taken by our much-missed friend and colleague Davy Murray. On 18 September 1993 Number 9 looks on from the shed yard at Markinch as the driver of A2 No. 60532 *Blue Peter* gives a friendly wave as he passes. *Blue Peter* was hauling The Fair Maid rail tour bound for Perth, which had started at Newcastle, with *Blue Peter* coming on at Carlisle for the run up the GSWR and across to Fife. (Davy Murray)

With classmate No. 4468 *Mallard* in the background, Number 9 stands in the sunshine at Doncaster Works' open day on 26 July 2003 (Davy Murray)

Number 9 crosses the Victoria Bridge on the Severn Valley Railway on a beautiful day in 1996. Victoria Bridge, when it opened in 1862, at 200 feet long, was the longest single-span cast-iron bridge in Britain. (Davy Murray)

Taken during the Severn Valley Railway 2002 autumn gala, Number 9 heads the railway's rake of LNER teak coaches past the safari park between Bewdley and Kidderminster. Number 9 was based at 'The Valley' from 1994 to 2007. (Davy Murray)

Putting a Pin in a Map

Bounds Green depot is a bloomin' noisy place at night – I'd hate to be a neighbour of theirs. Klaxons sound when there is movement in or out of the shed. Depending which end of the shed the move is determines which Klaxon sounds, that way staff, and neighbours, know at which end to be vigilant and it goes on for ages, stops and then starts again.

We had a few days at Bounds Green prepping Number 9 for a tour out of King's Cross on Saturday 20 July 2002 called The Lincolnshire Coast Express to Cleethorpes… Yes, hold that thought! We had two nights trying to sleep in the coach with the prepping work being carried out during the day. Needless to say, we didn't get much sleep!

In the early hours of the Saturday morning the train was made up and a Class 47 was to draw the train off the depot into King's Cross station. Johnny and I were chatting to the driver, and he offered us a cab ride into the famous East Coast Main Line station. On arrival, a little before 06.00, the passengers were already on the platform waiting for us. As Johnny and I climbed out of the Class 47 loco, we noticed a famous individual who was to be on the train that day. We spent a few moments speaking with Mr Alan Pegler, the man who saved A3 Pacific No. 60103 *Flying Scotsman*.

Our departure time was 06.20 and, for operational reasons, we found out we were no longer going to Cleethorpes but terminating short at Scunthorpe of all places. At least Cleethorpes has the sea whereas Scunthorpe has a steel works and is not quite set

Davy has placed The Elizabethan headboard onto Number 9 before taking this superb shot at Bounds Green depot. (Davy Murray)

up for tourists. Things were to change again as we set off along the hallowed former A4 stomping ground. The Lincolnshire Coast Express was now going to Yorkshire with Scarborough as the destination, so at least the passengers would be able to get a paddle after all.

We were booked inside the yard at Peterborough for water and during the stop I made tea and bacon rolls for the footplate crew and John Cameron, who was also on board. Unusually, our coach was at the back of the train, so I had to carry the rolls and tea on a tray, making sure I didn't trip on anything on the ground as I walked the length of the train. We were also booked to take water at the Royal Mail terminal just south of Doncaster. Unfortunately, our passage through 'Donny' station on the Down fast was not as fast as we would have liked. Having said that, the loco had several spells of sustained running in the 80s as we headed north. All this was of course before the days of black boxes and the like.

The rest of the trip to Scarborough was uneventful as was the start of the return trip to Doncaster, where we were booked to come off the train. As we approached the outskirts of York, we made slow progress and were held for quite a while. There was a thunderstorm to the north and this had affected the signalling, meaning the ECML was in a spot of turmoil. To add to the issues, Railtrack needed to put on an emergency track possession between York and Doncaster at a certain time. We eventually entered York station and drew up at the Up starter signal at the London end of Platform 3.

With barely a hint of exhaust, Number 9 speeds north between Arlesey and Langford on the East Coast Main Line with the north-bound Lincolnshire Coast Express. (Ian Bowskill)

Back in 1969 Alan Pegler took No. 4472 *Flying Scotsman* to America on the now famous Trade Tour. He also took a British Railways driver and two firemen, one of the latter being Dave Court.

On this day the driver who went to the signal telephone to speak with control was Dave Court. He was told, 'If you can do the 35 miles to Doncaster in 30 minutes, you can go.' He replied, 'Yes, we can', and off we went and how. Dave accelerated the train in fine style and with the high-speed track of the Selby Diversion between Colton Junction and Temple Hirst Junction where the original ECML is rejoined, we knew we were in for a good sprint. We covered the 35 miles in twenty-eight-and-a-half minutes with sustained high-speed running and a maximum of 87 mph and a hairy approach to Doncaster station. An A4 just did what an A4 was built for...running and running fast.

At Doncaster we came off the train and many of the passengers gathered around the loco and gave us a spirited round of applause. We retired to Doncaster Carr Loco and slept well that night.

Jules Hathaway

The Great Marquess – A Brief History

Locomotive No. 3442 emerged from the LNER locomotive works at Darlington in June 1938 and arrived in Fort William the following month carrying the incorrectly spelled name *Mac Cailien Mor*. Cailin was the founder and chief of the Clan Cambell,

The Great Marquess stands taking water at Crianlarich while working a Fort William to Glasgow service on 31 August 1938. (T. G. Hepburn/Rail Archive Stephenson)

not a popular clan in Lochaber and it was a puzzle to those in Fort William why this name had been chosen. It was even suggested that those down in Englandshire thought that it was Gaelic for 'The Great Marquess', the name the locomotive was supposed to have. The locomotive disappeared south and returned with its proper name, *The Great Marquess*, the following month, named after James Graham, 1st Marquess of Montrose.

It was the second of a class of six three-cylinder 2-6-0 locomotives built specifically to work the West Highland Line from Glasgow to Fort William and Mallaig and was based at Eastfield shed in Glasgow.

The design of these locomotives dates back to 1924 when there was a need for more powerful locomotives, to do away with the double-heading of the existing ex-NBR D34 Glen Class 4-4-0 locomotives. The use of the then new K3 2-6-0s was proposed, but they proved to be too heavy for the line, so as more K3s were produced, K2 2-6-0s were transferred from other parts of the LNER to the West Highland Line, proving their worth, being able to haul heavier trains than the D34s.

The use of K2s was a stop gap as a new design was proposed, but it wasn't until 1936 that the idea became reality when Sir Nigel Gresley started work on the design of a class specifically for working on the West Highland Line.

All, except for the first of the class, No. 3441 *Loch Long*, were named after clan chiefs, 'Mac Cailin Mor' even re-appearing on locomotive No. 3445.

Now carrying its BR Number 61994, *The Great Marquess* stands at Eastfield depot, Glasgow, awaiting its next duty sometime in the 1950s. (Rail Archive Stephenson)

The K4s proved themselves to be masters of the West Highland curves and inclines, being able to haul 300-ton trains unaided and with a lower water consumption than the K2s, but they did have an Achilles heel. Having driving wheels of 5 feet 2 inches, they didn't take kindly to the high speeds on the Craigendoran to Glasgow section of the West Highland or the level section along Loch Eil on the Mallaig extension, suffering from high levels of vibration and the slackening of the centre big end nuts, resulting in increased maintenance inspections.

With the new B1 4-6-0s and K1 2-6-0s being introduced in the late 1940s, followed by LMS Black 5 and BR Standard Class 5 4-6-0s later, *The Great Marquess* and the rest of the K4s were reduced to only working passenger trains on the Fort William to Mallaig line and goods trains over the rest of the West Highland before being transferred away to Thornton shed in Fife in 1959. Here, their high power was greatly appreciated by the Thornton crews while working the many coal trains in the area, in particular on the long climb up Glenfarg.

This wouldn't last long, however, as the entire class of K4s was withdrawn from service towards the end of 1961, *The Great Marquess* being the last to go in December and thankfully Thornton wasn't in any hurry to send it to the scrapyard.

This could have meant the end, but for the saviour in the shape of retired Scots Guards Major, honorary Colonel of The Fife and Forfar Yeomanry/Scottish Horse and railway enthusiast Viscount Lord Garnock, later Lord Lindsay and son of a director of the LNER, who purchased *The Great Marquess* in March 1963 from British Railways. It was then overhauled and repainted in LNER apple-green livery at Cowlairs Works in Glasgow.

The work was soon completed, and on 29 April 1963 *The Great Marquess* headed south down the East Coast Main Line to Leeds Neville Hill shed, which would be its base for the next nine years.

For the next four years, *The Great Marquess* worked rail tours all over England, but sadly never returned to Scotland at this time. One tour, on 3 October 1964, even involved double-heading with A3 Pacific No. 4472 *Flying Scotsman* between Harrogate and Darlington. During this period the locomotive was carrying a bell on the front buffer beam which had been presented by the Pennsylvanian Railroad, who also had K4 locomotives, only these were much larger Pacifics and, indeed, Sir Nigel Gresley had been influenced by these locomotives when designing his first Pacifics. In March 1967 *The Great Marquess* was the subject of a painting by the celebrated railway artist Terence Cunio, who painted the locomotive at Birmingham's Tyseley roundhouse flanked by Great Western Castle Class locomotives Nos 7029 *Clun Castle* and 4079 *Pendennis Castle*.

Following its overhaul at Cowlairs Works, *The Great Marquess* passes Cockburnspath on its delivery run to Leeds on 29 April 1963. It hauled this train of new Minfits as far as Heaton from where it hauled a goods train to Leeds and its new home. The wagon immediately behind the loco is carrying spares. (W. J. V. Anderson)

The Great Marquess double-heads The Whitby Moors rail tour with K1 2-6-0 No. 62005 at Market Weighton on 6 March 1965. The rail tour covered a number of now-closed lines in the area including Market Weighton. (53A models of Hull (D. K. Jones))

The Great Marquess was withdrawn with boiler problems in April 1967 and with the intended British Railways steam ban in 1968, Lord Garnock decided not to go ahead with repairs, and so the locomotive remained in store at Neville Hill until 1972. I observed it there in November 1968 along with ex-Great Eastern N7 0-6-0T No. 69621 and K1 2-6-0 No. 62005. The K1 had been retained as a source of a replacement boiler for *The Great Marquess*, but happily this never happened and the K1 made it into preservation under the care of NELPG. No. 69621 is now under the care of the East Anglian Railway Museum.

In 1972 *The Great Marquess* was moved to the Severn Valley Railway and boiler work was carried out at Bridgnorth, but it only ran for a short time on 'The Valley' before being stored pending a major overhaul. Lord Garnock, meanwhile, had become Chairman of the Severn Valley Railway at this time, succeeding Sir Gerald Nabarro. He would later become President.

By 1989 *The Great Marquess* had been overhauled and was recommissioned at Kidderminster Town station on 18 April by British Rail Chairman Sir Robert Read. In July of that year it returned briefly to the West Highland Railway where the by now Lord Lindsay, who was gravely ill, had a last footplate trip on his engine between Fort William and Banavie. He sadly passed away a few days later.

During the 1990s *The Great Marquess* as well as working on 'The Valley' and other heritage lines was kept busy on main line specials, including being the reserve locomotive for

The Great Marquess stands forlorn and awaiting overhaul at Bewdley on the Severn Valley Railway in June 1976. (Michael Mather)

Now fully overhauled to main line standard, *The Great Marquess* heads the Carlisle-bound Pennine Express at Dent Head, on the Settle to Carlisle Railway, on 9 July 1989. From Carlisle *The Marquess* and support coach headed north to Fort William. (Davy Murray)

the Forth Bridge Centenary special on 4 March 1990 should Number 9 hit any problems, although happily this didn't happen. It also took part, along with its former Neville Hill shed mate K1 No. 2005, in the West Highland Centenary celebrations in 1994.

For the Severn Valley Railway's autumn gala in 1997 *The Great Marquess* was repainted in British Railways lined black livery, for the first time in preservation. However, it wouldn't run long in this condition as a few weeks later it suffered a broken crank axle and was withdrawn pending repairs but was in a long queue of locomotives awaiting attention.

Following Lord Lindsay's death, the Lindsay family retained ownership of the locomotive, but in 2003 they sold it to John Cameron, who planned to restore it to main line condition. As there was no slot at 'The Valley' for this to be done, he moved it to Crewe Heritage Centre for the work to be carried out.

The story continues...

Michael Mather

We Worked on *The Marquess* Before

Long before Mr Cameron acquired *The Great Marquess*, what-with Number 9's and therefore our association with The Severn Valley Railway, the support crew occasionally helped out with the main line running of the loco. Here, fresh from overhaul, *The Great Marquess* stands in the shadow of Ben Nevis for the first time in thirty years, in July 1989. Lindsay was on the support crew on this occasion, and it was at this time at Fort William that Viscount Garnock had his final trip on the loco. (Davy Murray)

Right: Nick, Davy and an unknown SRPS volunteer giving *The Marquess* a clean prior to The West Highland Centenary rail tour in August 1994. (Michael Mather)

Below: Blowing off furiously, 3442 *The Great Marquess* and K1 2005 prepare to leave Crianlarich while working The West Highland Centenary special to Fort William on 7 August 1994. George was one of the traction inspectors on this train. (Marilyn Leven)

Working an SRPS Fife Circle rail tour, *The Great Marquess* passes the former Lumphinnans Junction on 14 September 1994. (Bill Roberton)

The Coming of *The Marquess*

Following its overhaul to main line standard at Crewe Heritage Centre, *The Marquess* moved to the Churnet Valley Railway in Staffordshire for running in and is seen here running alongside the River Churnet in January 2007. (Nigel Valentine)

Running in complete, the next stage was the main line loaded test run which *The Marquess* undertook on 8 March 2007, and makes a fine sight in this photograph taken near Hellifield during the test run. (Edward Osmotherley)

The next stage was the return to Scotland of Number 9 and *The Great Marquess* to work the inaugural Great Britain rail tour and to take up residence in their new home at Thornton. John and Mike pose for the camera while preparing the locos, on 5 April 2007, for the trip north. (Jules Hathaway, Michael Mather Collection)

A ghostly Fraser is shrouded in steam while stopped at Penrith during the journey north to Mossend yard on 6 April 2007. (Michael Mather)

Number 9 took over The Great Britain rail tour at Glasgow for the run to Perth where *The Great Marquess* was coupled on the front for the rest of that day's trip to Inverness. Number 9 makes a fine sight on a beautiful day approaching Bardrill Road, between Blackford and Gleneagles, on 10 April 2007. (Support Crew Collection and David Peaker)

Now with *The Great Marquess* on the front, John Cameron's two engines make an impressive sight at Dalnaspidal as they attack the climb to Druimuachdar summit, the highest main line railway summit in Britain, on 10 April 2007. (Bob Avery)

The following day *The Marquess* took the tour to Kyle of Lochalsh, a route to the West Coast of Scotland it would never have been on before. On arrival at 'The Kyle', the support crew pose for the camera, from left: Colin, Wesley, Steve Latham (Crewe Heritage Centre), Fraser, Jules, Marilyn and John. (Support Crew Collection)

The return trip to Perth on 14 April was done with Number 9 leading due to *The Marquess* having developed a leak at the superheater header. Coming off the train at Perth station, from where a Class 47 took it south, the two locos stand at Platform 6 taking water under the watchful eyes of Wesley and Marilyn before heading for Thornton. Ian, meanwhile, looks on from Platform 7. (Michael Mather)

Steam Revisits Thornton Junction

On a pleasant evening on Saturday 14 April 2007 the A4, K4 and support coach arrived somewhat late at the former EWS fuelling and stabling depot near Thornton, Fife, after working various legs of The Great Britain rail tour. So began a nine-year stay for these popular engines and here they would be maintained, prepped and cleaned for peripatetic charter work all over the UK. Being based at Thornton was very much welcomed by the support crew as most lived reasonably close to the depot. The facility itself was light years away from what we had at Markinch and included four buildings: a two-road shed with inspection pit, two storage sheds and an amenity bothy. This did mean, however, starting afresh in terms of the setting up and storage of all the tools, equipment, sundry spares, lighting, power tools and oils, and, some said most importantly, a much-needed bothy washing machine for dirty overalls, as I was banned from using our home washing machine! Over time, things began to take shape with storage racking, shelving and tool boards being installed in the main shed. Tools were now clearly visible, easily identified and a calming sense of organisation became evident. It wasn't quite the '5Ss of Lean Principles', but not far off it and it made life much easier. The bothy was where work requirements

Heads look out from cabs and doors as after a long day Number 9 and *The Great Marquess* arrive at their new home at Thornton yard on 14 April 2007. (Michael Mather)

would be discussed, scoped out, recorded and allocated to individuals. It was also a place of good camaraderie, stories, jokes and laughter. Creature comforts such as running hot water and kitchen facilities, including a cooker and fridge, were well used and much appreciated.

Both engines were stored in the two-road shed with the support coach kept outside on an adjacent road next to the shed. The coach windows were fitted with removeable window boards for security reasons as the yard was quite isolated. Moving a 'dead' loco out of the shed for steaming wasn't straight forward, and often one of John Cameron's farm tractors would be sent to do the needful as there wasn't a diesel shunter on hand. This was around a 40-mile trip for the driver, and I always wondered what the many vehicles queuing behind him on the narrow winding roads of the East Neuk of Fife thought of this tractor never seeming to go into any farm gate.

Another big bonus at Thornton was that coaling was quickly done by a telehandler, a far cry from the time-consuming and energy-sapping hand method in the early days at Markinch. Throughout the frequency of the ongoing tour commitments of the locos with running mileages mounting up, examinations, inspections and maintenance schedules had to be integrated within fairly tight timeslots. It also soon became clear that we lacked equipment in certain areas, especially turning and drilling capability as

John Cameron directs operations as the two locos reverse the support coach into its siding beside the loco shed. (Michael Mather)

well as oxy, fuel-burning, heating and welding equipment. We also needed a mobile mechanical lifting capacity that would clear both the height and width of the engines. Actually, the mention of the word 'heating' takes me back to the winters inside the shed. The building was basically a cladded steel structure which was very thermally inefficient and unheated – and in the darkest months of January and February it was cold enough to make your teeth pop and curl the toenails!

The first priority to arrive was a mobile gantry crane, which was timely indeed as the K4 had a leaking super-heater header joint at the tubeplate. Fraser had sourced the gantry from LNWR Crewe and re-assembling it in the restricted space we had wasn't as straightforward as you may think, but with a bit of ingenuity and some smart rope-work it was rebuilt and tested ready for immediate use on No. 61994. The chimney, petticoat, blast-pipe, steam pipes and access covers were removed from the smokebox to provide adequate space to work. A sturdy box section was then fitted to the smokebox ring to anchor a chain-pull to, then chain lashings were secured round the header itself. With the gantry crane taking the weight, the super-heater header assembly with the elements was carefully drawn off the mounting studs to gain access to the main steam pipe and header flange faces. A bespoke copper flange gasket was carefully placed between the mating faces, then the entire process done in reverse to form the steam tight joint. This was a time-consuming team effort which was well executed, with the K4 successfully passing its subsequent steam test.

Rain falls on the photographer as *The Great Marquess* stands over the pit and is framed by the mobile gantry on 16 August 2007, while being prepared for a trip. Note the giant Haig's Whisky bottle in the background, a remnant from our time at Markinch, the home of Haig's Whisky. (Jules Hathaway)

In 2008 the K4 underwent a small design modification at Thornton, with the installation of rear steam sanders to the centre driving-wheel set. This involved a pair of cannibalised sand boxes and piping from a scrapped Class 08 diesel shunter, which were then altered to suit the positioning on the main frames and their intended purpose. The addition of these, when necessary, greatly aided tender-first workings on adverse gradients on routes such as the West Highland and Blaenau Ffestiniog lines. Incidentally, the water capacity of the tender was increased by some 500 gallons around this time too, but this was completed during a scheduled visit to Crewe.

Next to arrive was a lathe, which I had personally bought from my employer at the time – it was a known entity to me and priced well below market value. The Dean Smith & Grace 1307 × 40-inch centre lathe came with enough tooling and equipment to get off to a flying start. It was to be housed in what we called the 'garage', a sturdy, secure outbuilding which, after some damp-proof lining was put on the walls and lighting, ambient heating and three-phase power were installed, would become a micro machine shop, including a pillar drill and pedestal grinder.

It was a dark, cold night when the hi-ab lorry arrived with the lathe. The yard lights illuminated the boardwalk as the driver reversed over two disused lines before lifting the machine off onto the slabbed walkway leading to the shed. About six of us were

Mr Cameron's tractor has arrived and hauled *The Marquess* out of the shed ready for the fire to be lit on 7 February 2009. (Nick Swierklanski)

braced to push the machine along the walkway, through the shed and on into the 'garage', a distance of around 100 yards or so. We were at the ready with machinery skates, pinch bars and pipes as the lorry drove off. Well, I can still hear the 'choice language' and remember the aches to this day! The 5,500-lb machine was then fitted with the skates at the corners – and we pushed, only for it to come to a sudden stop after a couple of feet! The walkway slabs were cursed and cursed again; they were so uneven that the timeworn skates just wouldn't work without jamming on protruding slabs. So, it was to be pipes, round bars and pinch bars all the way. An hour-and-a-half later the lathe was indeed in the garage, but to say we were all completely knackered and short tempered would be taking a liberty with understatements. Of course, soon after levelling off and wiring up, the machine was promptly put to work and was regularly used for the manufacture, repair and modification of many parts for both engines. It was an older lathe, but still in very good condition from a quality UK manufacturer. It is interesting to note that it was installed 'as new' in Dounreay Nuclear Power Station, then on to Fife College, then to Thornton and eventually found its way to what is now Locomotive Services Group at Crewe, where I believe it's still going strong.

All smiles from me while at work at the lathe. (Jules Hathaway)

Over the years the engines would be booked into Crewe, and latterly at Bury, for specialised examinations and associated work or, depending on the nature of what was needed, sometimes their boilersmiths/fitters would travel up to Thornton. We undertook as much of the work as we could within the skillsets available in the group. This could mean working on the engines when they were in steam and hot, out of steam or potentially during a charter. I recall on one such occasion when Number 9 was travelling light engine with the support coach from Thornton to Edinburgh Waverley via Craigentinny for The Great Britain VII and the exhaust steam injector had been playing up. After coupling up to the stock at Craigentinny and while waiting for our path into Waverley, Fraser asked me to crawl underneath and have a look at why the injector wasn't 'picking up' as it should. After finding where and what the fault was, I got underneath the cab again armed with ring spanners to see if tightening the water-feed to injector flange bolts would seal the leak. Alas, a quick test of the injector by Fraser on the footplate revealed that the damaged gasket was still passing water and would need to be completely replaced when the engine was 'cold'. Mike and I successfully split the joint and replaced the damaged gasket with a new one at Inverness the following day.

One of the more momentarily anxious moments for me was not of any 'hot workings' but actually during a 'cold' boiler inspection at Thornton during the winter months when the boiler was empty. All boiler washout plugs and mudhole doors had been removed including the two fusible plugs in the copper firebox crown. Fraser wanted me to re-form and clean out the threads and that meant a spot of hand tapping,

Number 9 stands at the back of the shed on 7 March 2009 having just had the fireman's side steam pipe replaced, which necessitated the removal of the chimney cowl and some of the streamlined casing to gain access. Sadly, we never envisaged writing books about our exploits, so photos of work being carried out at Thornton are few and far between. (Nick Swierklanski)

which wasn't a problem. The fly in the ointment, however, was that this would mean awkwardly standing stooped inside the firebox and tapping vertically into the soft crown sheet and the thread was BSPT, a tapered thread no less. So, ultra care was needed to re-engage the tap 100 per cent correctly into the existing thread pitch and not strip it out or damage it. I had a couple of lights in there with me to ensure excellent visibility. With the tap in the wrench, I had very carefully engaged the tapered tap into the original thread when Fraser's head lunged through the fire-hole saying, 'Mind and not double start them threads – that'll be a disaster' and didn't I just know it! Needless to say, albeit slightly spooked, it all went just fine, but a moment of check and re-check always pays off. Funnily enough, some might say that I wouldn't fit through the A4 fire-hole these days, and I might just have to agree with them!

Working on and around steam engines during preparation heightens the senses – the sight, sound, smell and heat of a live steam locomotive keeps you alert to the potential dangers. I always enjoyed working on an engine whilst in steam as it made the prior work seem all the more worthwhile. Ahead of any main line running, a 'Fitness to Run' inspection by a qualified, authorised examiner takes place. Set criteria for numerous function and visual inspection tests are carried out with the engine in steam twenty-four hours before going on the main line. Part of the FTR requires the engine to be moved around under power and I have had the pleasure of driving, under supervision of course, both the A4 and K4 around the yard, an amazing experience that I'll never forget and, oh, that feeling...

Most of my time with these locomotives was on the engineering/maintenance side of things, not so much on the operational side and I wouldn't have had it any other way. I enjoyed the variety of work which that involved and felt my career lent itself well to what was needed.

It was such a privilege to have been involved with the A4 and K4 in the many times of absolute triumph, and the occasional difficulty. It was also a privilege to have been the allocated 'duty fitter' on some of the many enjoyable main line runs and, of course, to have worked alongside other support-crew members, as well as other like-minded people within the wider main line steam scene. It was a joy indeed.

In closing, it must be said that due credit, respect and thanks must go to the owner, John Cameron. Without his innumerable years of steadfast commitment to main line steam locomotives, none of this would have ever been possible.

Note: The gantry crane referenced here was donated to the restoration project for LMS Ivatt Class 2 2-6-0 No. 46464.

Gordon McHattie

Taking *The Marquess* to the Fort

The support crew loved taking Mr Cameron's K4 *The Great Marquess* 61994 on the West Highland Line. It was what it was made for, taking the sharp bends and cruel inclines with ease. If we heard that a trip was arranged, there was great excitement. Fort William (the Fort) was the usual destination, of course, whether it was to be the motive power on The Jacobite or other adventures which I am happy to set out. Our shed at Thornton yard had two sets of rails, or roads as they are known in railwaying. The road with the pit was always used by the A4 *Union of South Africa* as it was longer than *The Marquess* and there was more room on that side of the building. To prepare for the run we needed to get the K4 out of the shed and over the pit, and so our regular shunter, Mr Cameron's tractor, would arrive from Balbuthie to do the necessary.

Once the K4 was outside the shed the fire could be lit in the firebox. There was only a single 2-inch hose from a stand pipe as our water supply, so it was practice to put the hose in the tender and let it run all night. The next day enough steam would be raised to allow the brakes to be let off and *The Marquess* could be used to shunt the A4 out of the back of the shed so it could take its place. This was a chance for the support crew to drive for a short bit. The yard was empty of any other vehicles or people so there would be no danger to anyone. Two sets of points needed to be shifted and the engine was in place. The shed had no system to expel smoke so it could only remain in until the necessary work was done.

By the next day the A4 was back in place, the support coach was attached to *The Great Marquess* and all oiling, coaling, watering and cleaning had been done. The driver and fireman for the trips were supplied by commercial companies, often West Coast Railways. A frequent driver in Scotland was Peter Walker and on one trip he decided that the support-crew member on the footplate should take a video film of the trip. I was handed the camera, which I had no idea how to operate, and I obediently

The support crew roll up the fire hoses at Crianlarich station, having just watered *The Marquess*, which is at the head of The Great Britain VI, on 27 April 2013. The loco hauled the train as far as Edinburgh, from where the following day, Number 9 hauled it to York. This was a very busy spell for the support crew, as before going to Fort William and Mallaig with *The Marquess*, they went to Aberdeen with Number 9 for it to haul The Great Britain V1 south to Edinburgh. (Chris Boyd)

pointed it at the workers on the footplate and the scenery going by. To this day I never did find out if I actually captured anything worthwhile.

Stopping at Crianlarich for water always proved a source of stories. Once someone had parked their car over the water hydrant. The owner couldn't be found, so the support crew bounced it off the hydrant. Names have been omitted to protect the guilty. Another time Mr Cameron was helping to clean the fire which had a bit of clinker in it. He was using the long-handled shovel called a dander shovel to remove the hot cinders out of the firebox and deposit them in the cess on the right-hand side. Unfortunately, this was where the signalling wires were, close to a low wall, and they duly started to smoulder. I had to turn on the injector water supply which flowed out the overflow pipe and could be diverted using a shovel onto the hot wires. No harm was done, thank goodness.

Facilities at Fort William were very primitive in the early days. Luxuries such as toilets and showers were provided by the good hearts of the ScotRail depot staff, who allowed us the use of their bothy. There was an engine shed but the roof had long gone, and the pit regularly filled up with water when it rained on Ben Nevis, which if you know Fort William, could be every day. I once cleaned out the ashpan over this pit after building a pile of pallets and donning the team wellies. There was only the one pair, and they were enormous, fitting anyone in the crew.

It always seemed to be nesting time for the herring gulls in the yard who used the roofs of the buildings for this purpose. In June darkness barely fell and that, combined with the constant mewing of the gulls, made it difficult to sleep. Even worse was when the gull young hatched as they would attack anyone moving about near them. Going for a shower required a hard hat and a stick.

The Scottish businessman Brian Souter hired *The Marquess* for a special outing for his AGM in May 2008. The party had been at Oban and travelled by rail to Crianlarich, where we took over having gone south from the Fort. We arrived safely at Fort William and discharged the happy passengers.

The northward journey to get to Fort William in the first place, double-heading with the 8F No. 48151, has gone down in infamy. From the support coach we could see hot cinders falling like rain from underneath the 8F's firebox and bouncing out from between the rails (the 4 foot) onto the tinder-dry vegetation at the side of the line. The inevitable happened and the countryside was ablaze.

The Great Marquess double-heads with 8F No. 48151 (the fire starter) on the climb away from Rannoch station on their way to Fort Willaim at the head of Brian Souter's special train on 6 May 2008. (Michael Mather)

Jules photographs Nick photographing *The Marquess* and 48151 crossing Rannoch Moor on 6 May 2008. (Jules Hathaway)

One of the support crew, Mike (the brains behind this book), who travelled on the service train from Rannoch that followed us, said that the fire was actually speeding as fast as the train at one point. It is no wonder steam trains can be stopped for fire risk. On the return trip, we stopped for water at Rannoch station and a water tanker was waiting for us along with six fire engines in the car park. Mr Cameron could be seen handing the firemen cash to buy a drink and thanking them for their help. The support crew was amazed by this, not because he was thanking them, as he was a great people person, but because he actually had money on him. He was known for being like members of the royal family, who famously never carry cash.

On 4 September 2010 we again double-headed with another loco to Fort William, this time the Black 5 No. 45407. It headed south the next day with a train for Oban and we followed on light engine and coach to Crianlarich. It was there we heard that the Black 5 had failed to make the steep climb out of Oban with its train and was delayed. Our support coach was shunted into the yard and the brave *Marquess* set off to the rescue with our Chief Engineer and Fraser on the footplate. They arrived back at Crianlarich, but everything was well late by now. Double-heading, we set off for Glasgow. The weather deteriorated as darkness fell and the footing was so bad at Glen Douglas that the support crew had to get out

A fine line-up of fire engines greets the train at Rannoch station on the return trip. They must have been quite a sight speeding their way along the narrow roads to Rannoch station. 6 May 2008. (Michael Mather)

and hand-sand the rails. We finally got into Queen Street at 04.30 and the station had to be opened specially. The passengers and the train catering staff gave the engine crews a standing ovation on the platform. It would have brought a tear to a glass eye, as they say in Glasgow.

This delay meant there was no chance to return to our home shed on the Sunday and the loco ended up in Caddor yard in Glasgow while another route could be planned. There were only three of us left, the rest of the crew having to make their own way home. Me, because my lovely boss allowed me a day's holiday at short notice, Fraser and our Chief Engineer, John Graham. At that time Caddor yard was a deserted backwater with no facilities at all, not even a water supply. When day broke there was just us, an occasional train passing on the main line some distance away over the empty lines and a solitary line of smoke rising off a smouldering sleeper left by the previous steam loco crew who must have cleaned out their ashpan. We eventually did get back to Thornton after an anxious day trying to keep the fire going and use only the minimum of the precious water in the tender.

Another special charter was for a group of railway photographers on Saturday 29 October 2011. It was an early start from Fort William up to Corrour, but it turned into an easy shift for us. The distances were not great so we didn't need to take water and could spend the day drinking tea in the support coach and watching the photographers out in the dire weather. Run pasts were executed by the crew so the engine could be seen at its best working hard. There were a lot of photographers climbing, being snagged on and falling over barbed wire fences, scaling sheer crags and shielding cameras from the rain. They were made of stern stuff, though, and some of the photographs taken that day were truly breathtaking.

Working on The Jacobite was a big job for us, meaning being away from home for a week at a time and for the volunteer support crew a hard slog if an enjoyable one. Up at 06.00 to prepare the loco then down to the station with the carriages for the train. The annoying thing was we couldn't take our own support coach so we had to take everything we would need for the engine with us. This included oil, hoses for watering and any tools we might require. We did get to use a compartment on the train, and it was a standing joke that as soon as the trip started the support crew would fall asleep. Arrival at Mallaig meant oiling up and taking water then shunting up the yard, where the engine would wait. Free then to go into town, we made a bee line for the chippy, where preferential rates for crew meant cheap fish and chips. Then a walk down the harbour to see the fishing boats and the ferry coming in from Armadale. One Spanish

On an unknown date, *The Great Marquess* makes a fine sight in glorious West Highland weather as it crosses Loch Eilt causeway on its way to Mallaig. (Support Crew Collection)

Colin, Jules, Marilyn and Fraser have a well-earned seat at Mallaig after arriving from Fort William with The Jacobite. (Support Crew Collection)

family was so delighted to see us all eating our chips from the paper that they had to take our photograph.

Back at the station, we would build up the fire and get the engine ready for the return leg. The turntable at Mallaig had long since gone so with no chance of turning the engine it was tender first on the return leg. A member of the support crew had to be on the footplate at all times, and this allowed us all to take turns. This was great going to Mallaig, but tender first was another story, especially when the weather was bad. Shelter from the wind and rain is scarce without the bulk of the boiler and the footplate spectacles. There was always a lot of, 'It was my turn last time and I've got the cold'.

Back at Fort William, the passengers were waved goodbye and then the engine would take the stock to the yard where the support crew could clean out the smokebox, firebox and ashpan. The engine was given a clean and the oiling started. The exhausted support crew could then go for a shower and get out for a meal before bed and then a six o'clock start the next morning.

Wonderful times and great memories.

Marilyn Leven

The Great Marquess stands at Mallaig ready to head back to Fort William tender first with The Jacobite. The Jacobite guard and good friend of the support crew Florence MacLean stands forward of the loco at the ground frame on 1 August 2010. (Peter Drummond)

West Highland grandeur at its finest as *The Great Marquess* takes The Jacobite across Glenfinnan viaduct on 1 August 2010. (Peter Drummond)

A Week in Steam

2007 turned out to be a very busy year. After the success of the very first Great Britain tours in April, Number 9 found itself with a hectic week of seven days' running towards the end of July. So, with annual leave sorted and bags packed, I set off for York where I would leave my car at the National Railway Museum. Then I took the train to Crewe where the loco was being prepared at the Crewe Heritage Centre.

Sunday 22 July was spent oiling round, cleaning and having the Fitness to Run exam and making sure everything was ready for the next day. We were in good company as the boys from Butterley were present working on their charge, Princess Coronation Class Pacific 6233 *Duchess of Sutherland*. The next day both locos would run together to York to position for a period of upcoming Scarborough Spa Express trips.

Monday 23 July, day one and we were all ready to go. The train was formed with Number 9 leading 6233 with its support coach and our coach at the end. Could this have been the first time a Duchess and A4 double-headed? Our route took us through

In what is probably a first, A4 Pacific No. 60009 *Union of South Africa* and Princess Coronation Pacific No. 6233 *Duchess of Sutherland*, double-heading, depart Crewe with their support coaches on their way to York on 23 July 2007. (Jules Hathaway)

Number 9 stands at York station waiting to depart for Scarborough with The Scarborough Spa Express on 25 July 2007. I risked being left behind as I crossed the footbridge to grab a couple of shots before returning to the safety of the support coach. (Jules Hathaway)

Stoke on Trent and Derby and on to Chesterfield. We stopped for water near Barrow Hill at a spot we named 'Dog Mess Junction', having once laid our hoses to the hydrant through a grassy area used by dog walkers. We actually called it something else but I'm sure you get the drift!

On arrival at the NRM we split the train with the *Duchess* and coach going to one siding and Number 9 going over the pit in their prep shed. We set about cleaning the fire and emptying the smokebox and ashpan. We were joined by two old friends of mine from the Swanage Railway, Steve Dyer and Pete Sykes. We were all set.

Tuesday 24 July, day two and the first of three Scarborough Spa Express tours. We dropped through York station to Holgate sidings to pick up the stock then returned to the station. The route for these SSE tours was York to Harrogate and on to Leeds, then back to York and the Scarborough branch to the seaside terminus. The return journey was the same route in reverse. At Scarborough the loco was turned on the turntable and at York it turned using the triangle just outside the station. Then we returned to the NRM to dispose the loco and prepare it for the next day.

Wednesday and Thursday 25 and 26 July, days three and four, and the trips went well with great runs and well-loaded trains. The support crew and loco crews were having a great time, but the work was hard. We had to eat all our meals in the coach, and we took it in turns to cook the evening meal. There was one incident when RO Fraser stood too close to the firebox door and the heat caused the coins in his jeans to heat up and badly scold his leg. Marilyn needed to look at his injury and she shouted to him from one end of the coach to the other, 'Fraser, come here and get your trousers off!!' I was asked to take photos...scary stuff! Once we arrived back at the NRM he was taken up to York Hospital to have it looked at and treated. As a result of this, Marilyn turned head nurse and banned him from the footplate for the rest of the week! Steve and Pete left us to return to Swanage, cheers, guys.

Friday 27 July, day five, saw us run engine and coach from York to Bury on the East Lancs Railway to position for a tour from Manchester to Carlisle and back the next day. We were routed south through Castleford and Wakefield, passing Healey Mills yard and on to Huddersfield. From there we headed for the Pennines to pass through Standedge tunnel and then to Manchester and Castleton where the East Lancs Railway has its main line connection.

On arrival on the ELR metals, we said goodbye to the West Coast loco crew and due to Fraser's ban from the footplate. I was nominated to drive the loco with Marilyn firing and an ELR pilot man on board. On arrival at Heywood station, there was a service train in the station and the plan was for us to couple on the back for the few miles to Bury station. The service train was headed by ex-GWR 37xx 4-4-0 No. 3442 *City of Truro*. At Bury we tied off and dropped down to the loco shed to dispose and prepare for the next day's tour.

In what is probably another first, Number 9 and GWR 4-4-0 No. 3440 *City of Truro* share a train on the East Lancs Railway, 28 July 2007. (Nigel Valentine)

Returning south from Carlisle, the usual water stop was taken at Appleby and Fraser and myself, with synchronised legs, and Marilyn behind, take a break on Lindsay's seat. (Jules Hathaway Collection)

Saturday 28 July, day six, and a rude awakening at 3 o'clock in the morning heralded a potentially good day. Whilst most of the oiling had been done the night before, one thing that has to be dealt with at the time is the air pump. This is situated between the frames just in front of the firebox. To reach it involves contorting the body whilst standing on the brake rigging, which is good fun to do from a pit and it is even more fun when there is no pit! You have to manoeuvre between the centre and rear driving wheels on your back then sit up and then get to your feet and turn to face the pump. What made it even more fun this morning was being underneath and told to come out as the loco had to be moved. You can imagine my sheer delight at being underneath a second time and again being told to come out. Coarse language ensued! After another move it was third-time lucky and the job was done, and I was filthy. At 05.15 we set off for the run to Manchester Victoria to pick up our train.

The tour took us to Carlisle via the West Coast Main Line over Shap and back via the Settle and Carlisle line over Ais Gill. Our route from Manchester Victoria was through Pendleton and Atherton to join the WCML at Wigan. We stopped for water at Carnforth before attacking the climbs to Grayrigg and Shap. At Grayrigg I went forward to the footplate and as we passed through the Lune Gorge, the fireman needed a break so handed me the shovel for the ascent to Shap summit. No pressure then...well actually a lot of pressure as once we were over the top and the regulator was closed the

Number 9 makes a fine sight as it crosses Ais Gill viaduct with The Cumbrian Mountain Express on 28 July 2007. (Simon Lathlane)

loco blew off. On arrival at Carlisle, we turned on the triangle south of the station and serviced the loco, taking on water and coal for the return trip.

The return saw us going via the Midland Railway's classic route over Ais Gill and we had a superb run south. Water was taken at Appleby and during the stop Marilyn, Fraser and I rested for a short while on the seat we installed there in memory of our dear friend, Lindsay Spittal (see our first book, *60009 Union of South Africa: Stories from the Support Crew*). At Hellifield we bore right for Blackburn and another water stop. From Blackburn we went left heading for Darwin and Bolton with journey's end at Manchester Victoria. We then had the short run back to Bury for another disposal and prep for the next day, which was not far away.

Sunday 29 July, day seven. Our last day, and we headed back to York not engine and coach, but we took the stock from the previous day and a dead Class 47 on the back. It was an uneventful run which went via Standedge tunnel and Huddersfield, but at Wakefield Kirkgate station we went right then took a left at Croft West Junction going through Featherstone, Pontefract and Ferrybridge. At York we left the stock and Class 47 in Holgate sidings and slipped back into the NRM sidings to allow the fire to go out and Number 9 to cool down.

It was one heck of a week with a lot of hard work done by the support crew and the loco, which clocked up an impressive 1,224 miles of fault-free running.

Jules Hathaway

Thornton Visitors

Over the years, we hosted several visiting steam locomotives laying over at Thornton between rail tours. The first to visit was new build A1 Pacific No. 60163 *Tornado* in March 2009, seen here dangerously close to the acetylene cutter! (Michael Mather)

In April 2010 Princess Royal Pacific No. 6201 *Princess Elizabeth* visited between Great Britain III duties and makes a fine sight outside the shed. (Jules Hathaway)

BR Class 7 Britannia Pacific No. 70013 *Oliver Cromwell*, the last Pacific locomotive that British Railways operated, stands inside the shed on 20 March 2012. (Michael Mather)

The Great Marquess and A4 No. 4464 *Bittern* stand outside the shed on 21 May 2012. Both locos had been involved in The Cathedrals Explorer rail tour. *Bittern* had brought the train up to Perth from Durham and *The Marquess* was involved in the Inverness to Kyle of Lochalsh section, with *Bittern* taking the train south a few days later. (Jules Hathaway)

The Opening of the Alloa Line

The support crew was always very pleased to have the chance to take part in the opening of a reinstated line in Scotland. When Mr Cameron let us know that No. 61994 *The Great Marquess* was to be heading a train at the opening day of the line from Stirling to Alloa on 15 May 2008, everyone wanted to be there.

West Coast was the train company in charge and the SRPS coaching stock hauled by a Class 55 Deltic No. 55022 *Royal Scots Grey* arrived at Thornton shed the day before. We had *The Marquess* gleaming by then as it was close to home so there were plenty of willing bodies to do the work. The diesel was looking a bit dirty, though, so Nick and I took to cleaning the front end to smarten it up a bit. We had never had a Deltic in the yard before and it was attracting a great deal of interest.

The day of the event started early with driver Peter Walker on *The Marquess* and Gordon Sellers on the Deltic. *The Great Marquess* coupled up to the support coach and then onto the coaching stock and we made our way to Alloa via Dunfermline with *The Marquess* leading and the Deltic on the back of the train. This meant that *The Great Marquess* would head the train from Alloa to Stirling, with the Deltic heading the train back to Alloa.

There were four round trips on the day, the first one for VIPs. The next three were for paying passengers.

A busy scene at Thornton yard on the opening day of the Alloa Line sees *The Great Marquess* coupled to its train and awaiting *Royal Scots Grey* to couple onto the other end. Deltic driver Gordon Sellers walks towards the camera and in front of Number 9 are John Cameron, left, and *The Marquess*'s' driver Peter Walker, right. (Michael Mather)

The view from the support coach of the new Alloa station and the crowd on the over bridge. (Brian Read)

The Great Marquess prepares to leave Alloa for Stirling with one of the opening specials. (Jim Ramsay)

Mr Cameron was in his usual fireman's seat the whole day and waved and blew the whistle for the onlookers. A large group of school pupils and their teacher were lined up along the railings of a school playground when we passed by. A treasured memory for them, I am sure, of their school days.

The platforms at both Alloa and Stirling were full of well-wishers wanting to take photographs of the loco and its owner.

Nick had asked permission to have a cab ride in the Deltic, which he enjoyed very much even though he had to take a lot of good-natured ribbing from the hardened steam fans in the group.

The first day of the new line's official timetable was Monday 19 May but no steam, of course, just the usual trains.

Marilyn Leven

The Highland Main Line

The Highland Main Line wasn't somewhere Number 9, or *The Great Marquess*, ventured on to when in service, but both have made several runs on it in preservation including this SRPS rail tour which Number 9 worked from Perth to Inverness and return on 4 July 2009. Here, near to Dalnaspidal, Number 9 is digging in to the 1 in 70 climb to Druimuachdar summit. (Michael Mather)

Heading south now on the return run, Number 9 is about to cross the Findhorn viaduct at Tomatin and will soon attack the 1 in 60 climb to Slochd summit, not something A4s were built for, but manage very well nonetheless. (Michael Mather)

While on an engine and coach trip to Inverness in readiness to work 'The Great Britain' to Kyle of Lochalsh, on 11 April 2010 *The Great Marquess* has made an impromptu stop so that Fraser can walk forward to retrieve his mobile phone, which he had inadvertently left on the loco's running plate. Such is the steadiness of *The Marquess*, it hadn't fallen off. (Brian Read)

Mike and John Cameron wave to Jules as *The Great Marquess* passes Druimuachdar summit while working an SRPS rail tour to the Speyside Railway at Aviemore on 26 June 2010. (Jules Hathaway)

The Mid-Norfolk Railway

Taking both of Cameron's engines out at the same time was always a strain on the support crew and it was with thanks to our many friends and helpers in all parts of the country that we managed to do as much as was asked of us.

Taking both engines to the Mid-Norfolk gala on 24 June 2014 was no different. We were delighted to go, of course, but a mid-week engine movement from York to Norfolk to get there the day before meant there were only four of us to prep, travel and attend the gala with both *Union of South Africa* (Number 9) and *The Great Marquess*: me, Fraser, David and John. Number 9 had been out doing a Cathedral's Express from Ely and *The Marquess* had been at the East Lancs Railway as a guest engine prior to them both ending up at York Railway Museum. Luckily, they were both in steam so all that was needed was taking water and oiling up prior to the daily Fitness to Run exam and we were ready to go.

The worry was that Number 9 wouldn't get any coal at York and Fraser was very concerned that we would not have enough for the journey. We were to get filled up at Dereham on the MNR and it would be a light engine movement to get there, so no heavy working to worry about. The coal stored at York belonged to West Coast Railways and there had been no prior arrangement for coal to be paid for, so we were stuck.

It was a beautiful day for the trip and there were a lot of photographers out to see the double-header pass. We arrived at Wymondham, where the main line connection joined the Mid-Norfolk rails, in pouring rain, though, and the main line crew left to get transport back to London. We quickly learned from the railway volunteers that we were to drive the engines up the line ourselves, which wasn't an issue as both Fraser and David were capable drivers and John and I could fire for them. Fraser and I were on Number 9, which by then had not a speck of coal left in the tender. Fraser was muttering oaths about the reason we hadn't been given more coal at York, but all the local railway staff could do for us was suggest that they had recently cut down trees by the trackside and the sawn logs were stacked next to us. A4s were never designed to run on wood, although Number 9 had done it once before at Inverkeithing, so it was worth a go. I discovered that the logs were covered in beetles and snails and the like, and I was trying to save their lives as Fraser made me sacrifice them to the flames with a 'Get the fire made up!' He wasn't in the best of moods by this time. Luckily, there wasn't a fire risk trackside as we threw out some spectacular sparks even going slowly. We passed through stations I had never heard of: Wymondham Abbey, Thuxton and Yaxham and other delightful halts. It is a lovely line.

On a beautiful summer's day, *The Great Marquess* heads south along the Mid-Norfolk Railway near Danemoor summit while on its way to Wymondham. (Ian McDonald)

Safe arrival at the other end at Dereham was a relief and we were made very welcome by everyone, especially Marie and George Saville. George was the general manager at the time and Marie was an accomplished driver and fireman.

It was wonderful to meet volunteers who were really pleased to be cleaning the visiting engines, and there was no shortage of help with everything. We even got the use of a washing machine which maybe doesn't have a relevance for the locos, but for a weary support crew who had been on the road for over a week by this time, it was a godsend. Friday gave us a chance to catch up with minor repairs and get both locos into top condition for showing off. The local volunteers were delighted to be shown how to oil up and were fascinated by the corridor tender and the power of an A4 compared with the little engines they were used to.

The gala was over Saturday and Sunday, and it was very busy given the hot, sunny weather. There was no turntable, so the engines were running tender first from Wymondham Abbey to Dereham, then facing the right way on the return leg after running round the train. The four of us took turns at being the owner's representative on the engines, which meant we each had a half day off to relax and enjoy the gala. Duty on the engines brought a lot of responsibility because the driver and firemen on duty were maybe not familiar with the engine, although they all did a really good job. Having

With John looking forward, Number 9, running tender first, passes Wymondham Abbey with a train for Dereham. (Ian McDonald)

Marie and Marilyn look down from the cab of *The Great Marquess* while in Wymondham Abbey Loop. (Ian Mcdonald)

a woman in a position of being in charge was still a novelty to some of the gala visitors and I was very pleased to be applauded by the many women on the station platforms.

I did hear that one elderly gentleman had asked in high dudgeon who the woman was on the footplate. When he was told I was the Responsible Officer (and for a A4 at that) he had shuffled off most put out. It is hoped he has managed to ease himself into the twenty-first century by now.

There was a water column on the north end of the platform at Dereham and it was always a delight to be able to use this piece of living history to fill the tender. Main line running these days means water tankers at stations for the most part, which doesn't have the same romance. I did soak an unfortunate man who wasn't listening to my shouts to stay clear when I swung the column arm over, but a chance to come up on the footplate to dry off more than compensated him for a wet shirt.

On the Sunday afternoon Mr Cameron arrived. We had not been sure he would make it because it is quite a trek from his farm in Fife to Dereham. He took up his favourite position in the fireman's seat of Number 9 and waved and blew the whistle to the delight of the visitors. He always loved talking to those admiring his engines and he beamed all afternoon.

A4s were built for speed on the East Coast Main Line, and they do not like going slowly. The 25-mph restriction on preserved lines means they never really get into their stride and the

The two locos stand in Dereham yard and with Fraser looking on, John wields the spanners on Number 9. (Ian McDonald)

high temperatures over the weekend heated up the water in the tender which caused problems with getting the injectors to lift. The fire was well clinkered up as well one lunchtime when I arrived on the footplate, and the engine was losing steam. The driver was most amused when I grabbed a poker and set to loosening the accretion on the fire bars under the door, immediately allowing better combustion and the steam pressure rose as we looked. It was something I had done many times before, but he had never worked with me.

The lovely people working on the railway took us out for memorable meals at local hostelries, and this was much appreciated. Leaving on the Monday morning was sad, but we had had a great time as well as swelling the coffers of the railway. I got to drive *The Great Marquess* from Dereham to Wymondham in our double-header formation on the way out, which was an added bonus. Opportunities to drive other than a bit of shunting in our yard at Thornton were scarce. There were locals and volunteers at all the level crossings who cheered and waved as we passed, and the main line crews were waiting for us when we arrived at the main line connection. Back to reality but another jewel in the memory box.

Marilyn Leven

Under Repair

On the 18 January 2008 Number 9 paid a visit to the Thornaby wheel lathe for tyre turning, the last loco to be so treated there as the lathe was removed soon after. The driver's side leading wheelset is seen here mid-cut in the tyre turning process. (Jules Hathaway)

In June 2008 we made use of the wheel drop at the National Railway Museum to remove the Cartazzi wheelset from Number 9 so that work could be carried out on the journals. In this view, the wheelset has been dropped down into the pit watched by engineer John Graham who is operating the lift. (Jules Hathaway)

Fraser casts an eye on the wheelset as, complete with axle boxes, it is lifted clear before being loaded onto a lorry to be taken to Crewe Heritage Centre for attention. (Jules Hathaway)

With the driving wheels removed, Number 9 stands at Bury during axle-box repairs on 5 April 2017. (Ian McDonald)

Getting Number 9 Ready for a Run

One of the frequently asked questions when the Number 9 support crew are able to talk to the public is, 'How long does it take to get the engine ready'? To answer that question, I have set out this story, based on my own recollections of preparing the loco in a depot that was well used in the later years of service. All the work was done by unpaid volunteers apart from our engineer Fraser, who is an independent contractor.

Two days before a main line run, which in this case was to be on a Thursday, saw only Fraser and me arriving at Southall depot in the early afternoon. Mid-week runs are often poorly staffed due to support-crew work commitments. The engine was in the shed, which was great because it was coming down stair rods outside. I had travelled from Fife via Waverley station and King's Cross and was very wet from having to walk from Southall station into the depot.

We stowed our gear in the support coach and changed into working togs. The electrical connection to the coach that had been put out when the engine arrived on shed more than a week ago was still attached, so we had the luxury of the compartments being warm and dry. The fridge was also working so there was milk for tea.

Once on the footplate, Fraser checked the water level in the boiler which was showing no pressure. It was down to a quarter of a glass, possibly due to a slight blow through on the right-hand injector while it was cooling. He set about disconnecting the left-hand water gauge to provide a handy port to put water in without removing a wash-out plug.

I opened the damper on the ashpan and removed the spark arrestor from the front of the pan from underneath. I then put a hose in the firebox to wet down the considerable ash and clinker left after the last run, then climbed in after putting on safety goggles and a mask and taking a lamp, shovel, brush and pry bar in with me. My goggles immediately steamed up due to the high humidity, but there wasn't much to see anyway until the ash was cleared. Digging down exposed the fire bars, which I lifted and put aside until I had a space down to the ashpan beneath. Once I was able to stand up in the ashpan the job became much easier, pushing and shoving the deposits underneath me then cleaning up the grate before replacing all the bars.

Fraser had started to clean out the ashpan from the pit, and he caught my boots a couple of times with the rake as we jockeyed for space. After a quick check of the tubes and crown stays in the firebox, I climbed out and began to set the fire: forty shovelfuls of coal just behind the firebox door, a broken-up pallet on top then a bucketful of oily rags on top of that. A few rags soaked in diesel were placed on the top and the fire started first match. Before I lit the fire Fraser had opened the cod's mouth to access the smokebox, which did not need cleaning out as it had been converted to self-cleaning some time ago, although it still needed to be inspected. I made sure the damper was closed and the drain cocks open before I left the fire to get going and went to oil up underneath.

Fraser had finished replacing the spark arrester inside the damper in front of the ashpan and was out of the way underneath by this time, so I started at the back by the air pump and worked my way forward doing driving boxes, big end, wee end and bogie boxes in turn.

After a general inspection of the engine, we filled the tender with water and Fraser attended to a few minor repairs. I put the headlamps for the run on charge then made a quick inspection of the tender oil trays by drawing them out and checking for water, inspecting the brasses and pads then filling them up.

Once a bit more coal was put on the fire it was time for a wash and dinner. Two more of the support crew were turning up in the morning so we were able to relax a little knowing there would be more help the next day. We headed out to a local bar-café for a meal, stopping off at a large supermarket to replenish the support-coach stores on the way back. On shed we banked the fire up to keep it going till the morning before turning in.

I was up at 6.30 the next morning and immediately went to check the water level in the boiler. It was fine and 10 lb was showing on the pressure gauge. This is enough to get the blower on if required. I put more coal on then went and put the all-important kettle on. The boiler pressure needed to be brought up to at least 200 lb for the daily Fitness to Run inspection. One of the support crew due to arrive was an experienced fireman so he would be able to take over building up the fire.

I finished oiling up all the outside oiling points which are called 'the corks' and filled the mechanical lubricators with cylinder oil in the leading box and lubricating oil in the trailing one. Everyone usually helps turning the lubricator mechanism to get oil going into the system. It is hard on the arms but a necessary job as the delivery pipes can be fairly empty if the engine sits for a while.

Here I am cleaning the rods at Bridgnorth on the Severn Valley Railway. (Michael Mather Collection)

There was a great deal of work to do before the engine could be moved out of the shed to get coal at the coal pile. The support crew are mostly able to work without instruction after many years' experience, and everyone set about the required jobs.

Two of us cleaned the engine and tender, the cab and the support coach and the others adjusted the brakes on engine and coach and the other many small jobs requiring attention.

Fraser was doing the Fitness to Run exam and could be heard tapping with his small hammer checking every nut, bolt and mechanism.

At last, the boiler pressure was high enough to get the injectors on. Both were tested to make sure they were lifting cleanly and shutting down again afterwards. Once 200 lb was achieved on the pressure gauge the air pump was turned on and the engine was reversed under its own power out of the shed to take coal. When the run is to be very long it can be necessary to fill the footplate with a couple of feet of coal, which will be used to keep the fire going while stationary in the shed and save the tender stocks for the run. This time it wasn't necessary.

While the engine was out of the shed, two of us cleared the ash from the pit into a skip. Not the most popular job but it is the only opportunity. Being out of the shed is also convenient for testing the safety valves on top of the boiler. The steam pressure coming out of these can damage shed roofs. Back over the pit again, the electrical leads were plugged in to charge up the batteries for the cab lighting, the radio and the AWS/TPWS on the engine. The manual and steam sanders were tested, as was the AWS /TPWS. This requires a magnet to be passed under the receiver underneath at the front of the engine, simulating the signal picked up from the apparatus between the rails while on the go.

The coach water tanks were then filled up at each end to serve the toilet and kitchen while on the move and all the safety checks completed on the coach.

Fraser had picked up a few minor repairs during the Fitness to Run inspection that were needing to be done and they were put to rights by the able engineers on the crew.

All the checks on the air braking system were completed and Fraser disappeared into his compartment in the coach to complete all the paperwork for the safety checks and produce the Responsible Officer's paperwork for the next day. The Responsible Officer is a member of the support crew on the day. It is always someone with experience of the running of the loco, though not always an engineer.

By late afternoon we were all tired, dirty and hungry and we started to take turns using the one shower cubicle available. Leaving the engine with a blacked-out fire just ticking over, we walked out to the main street and to dinner in a favourite Indian restaurant. It was to be a very early rise the next morning as we were due off shed at 05.30. When we returned the fire was stoked for the night and the boiler filled with water. All the fire irons and tools were stored away and everyone hit their bunks before 22.00.

I was awake just before my alarm went off at 03.30. I could hear everyone's alarms going off as well and some footsteps passed my door, probably someone going to check on the fire. It was cold and dark outside, and everyone got on with the work quietly and efficiently. The fire was built up and the tender topped up with water. I filled the syphon boxes on the sides of the boiler under the casing hatches with oil and finished the oiling on the tender brake rigging, tender axles and in the cab, topping up the air-pump lubricators so it could be turned on.

The AWS was switched on and the radio initialised. The cab lighting was working, and the Responsible Officer made sure all safety equipment was in the cab. The engine lamps and headboard were put on the front of the loco and all ladders and tools were stowed.

The electrical supply to the cab was removed and the cables stowed away. The coach electrics would be left on until 15 minutes before the off. The hose was disconnected from the tender once it had its last top up and rolled and stowed away.

Once all the support crew were able to return to the coach, they signed the paperwork for the day with details of their PTS numbers and acknowledging the safety briefing they had been given by the Responsible Officer for the day.

At 05.00 the crew turned up. The engine was completely ready to go: a good live fire in the grate, just over 200 lb on the pressure gauge and every check done. The crew was from West Coast Railways and both the driver and fireman were well known to us and greeted us as old friends. Tea was brewed and promises made for bacon rolls as soon

At Southall depot on 15 February 2019, the day before working a rail tour to Worcester, Jason turns the crank handle to close the 'Cods Mouth' having just been checking that all is well in the smokebox. As can be seen, we have still to complete the cleaning of the front end. (Michael Mather)

In the early hours of 7 March 2020 Number 9 stands in Southall shed, all prepared and ready to work its last main line rail tour, which it will haul from Ealing Broadway to York. (Michael Mather)

as we were at the station. The driver signed the sheet for the journey, checking all the support-crew details.

The RO was up in the cab for the journey to Victoria station which would take a couple of hours. Everyone else had a bite of breakfast and went for a short sleep. We were all up in plenty of time to change into smart bib and brace and jacket (blues as they are known) and make the bacon rolls and, of course, more tea.

Once on the station this would be the first time the passengers and public would catch sight of the support crew, and many would not have a clue what we had done over the previous two days to ensure the engine was ready for another trip.

The engine would not be on shed again till after midnight. We were generously given expenses by John Cameron, the engine owner, to ensure we could travel to the engine and have decent meals while we were there, but we did not get paid for the long hours we put in. We did it because we loved being able to bring joy to everyone who loves steam engines, and we had a thoroughly good time doing it.

Do I miss it all now the engine is retired? Not all of it. I don't miss getting up at silly o'clock on bitter mornings. I don't miss the occasionally primitive conditions we have had to live in when the engine and coach were stabled in obscure sidings without power or toilet facilities. I would go back and do it all again, though, because it was the best adventure ever.

Marilyn Leven

The Great Marquess on Heritage Railways

The Great Marquess has been a great favourite on Britain's heritage railways, especially on those with steep gradients where its power could be used to good effect, as here on the North Yorkshire Moors Railway. On 20 October 2007 *The Marquess* and K1 No. 62005 are playing to the gallery as they climb the 1 in 178 away from Goathland with a recreation of The Whitby Moors rail tour, which ran on 6 March 1965. (Jules Hathaway)

Silhouetted against the setting sun and reflected in the water, *The Great Marquess* hauls a goods train past Burrs on the East Lancs Railway on 4 December 2008. (Ian McDonald)

The Great Marquess stands at Boat of Garten station on the Speyside Railway at the head of an SRPS rail tour that is about to return to Perth, beside Caledonian Railway 0-6-0 No. 828 which is working a Speyside service, on 26 June 2010. (Jules Hathaway)

The Great Marquess pilots WD 2-8-0 No. 90773 between Haworth and Oxenhope on the Keighley and Worth Valley Railway during their spring gala on 9 March 2014. No. 90773, which was repatriated from Sweden in 1973, is the only member of its class in Britain. (Michael Mather)

Going Home

My journey to the support coach started on Saturday 14 February 1976, just a week before my fourteenth birthday. The Southern Steam Trust had finally been given the keys to Swanage station as the rebuilding of the Swanage branch was starting to pick up. After that day I returned in 1979 at a more appropriate age and spent four years as a volunteer with the track gang and loco department. I was the diesel second man on the very first train to carry fare-paying passengers in August 1979.

I joined the Number 9 support crew in January 1990 after the RAF brought me to Fife and I crossed tracks with Mike Mather, but I have always maintained my connection with the Swanage Railway. When the loco had a tour to Swanage for the first time, that was one I just had to do. My son Ross had also volunteered at Swanage and joined the support crew, so he put his name down for this trip too. The tour was a Thursday one from London with a second tour one-way from London to York two days later, which we would do as well.

The loco was stabled at the former GWR loco shed at Southall, West London, so Ross and I made our way there by train to meet up with the others.

The trip was on Thursday 21 September 2017, and we had an early start as we had to travel empty stock, with No. 57314 hauling to London Victoria station via Willesden. The diesel would stay on the back as it was needed to haul the train from Swanage to

Southampton on the return run and for the ECS move from Victoria back to Southall. Our departure time was 08.43 from Platform 2 and we arrived a good 30–45 minutes before hand. Passengers and curious commuters came over to have a look at the loco at the end of the platform. One guy was due at his work shortly and I tried my hardest to persuade him to 'throw a sickie' and come with us for the trip. My old pal from school and a senior Swanage driver, Pete Duncalfe, was tempted but he stayed for a short while before heading off to his work.

We left on time, departing the former London, Brighton and South Coast Railway and London, Dover and Chatham Railway station and crossed the River Thames onto the south side. We passed between two iconic landmarks, the former Battersea Power Station and the current Battersea Dogs and Cats Home and took the line for Clapham Junction to join the former London and South Western Railway route to Basingstoke, Southampton, Bournemouth and Weymouth. Two water stops were made at Winchfield and Beaulieu Road.

At Wareham we drew up at the end of the Down platform where our conductor driver was waiting with the token to Norden Gates. There were a few people on the platform to photograph the train and standing with Steve Dyer, the Swanage conductor, was the station booking clerk who had closed the office for a short time. Once we stopped, I left the coach and went over to hug the booking clerk, my eldest brother Danny, he who got me interested in railways, and other things, as we grew up in Wareham. I said hello to

A first for Number 9 as it runs into Swanage's Platform 2. (Andrew P. M. Wright)

Here, I step down from the cab onto the platform at Swanage, a place dear to my heart. This has been a long time coming.' (Andrew P. M. Wright)

Steve, another old friend and former occasional support crew member, and we boarded the cab with me taking over from John as owner's rep for the run down to Swanage.

The West Coast driver got out of the seat for Steve to take over and we set off with Worgret Junction just over a mile away and the divergence onto the branch. I have ridden the full length of the branch in a diesel once and numerous times on both steam and diesel locos on the heritage section between Norden and Swanage, but this was something exceptional and quite emotional.

Our arrival at Swanage was met by a sea of people including Swanage Railway press manager and friend Andrew Wright. We came to a stop just short of the stop block and I finally fulfilled a dream I had had for the last twenty-seven years, to step off the footplate onto the platform there. Amongst the people waiting was my sister Joanne and my late dad Doug. Ross joined us from the coach and Andrew set about getting the photos he wanted. During the three-and-a-half-hour layover, the train was shunted and watered and the loco went onto the shed for servicing. At 57 feet the shed turntable is too short to turn the loco, so No. 57314 would haul the train back to Southampton.

We departed at 17.25 with Ross on the footplate, where he stayed until we passed through Poole. Dad was waiting for us at Wareham to wave as we slowly passed through, stopping briefly for Steve Dyer to leave us and return the token into the equipment. The run to Southampton was uneventful and on arrival we paused for fifteen minutes. On leaving, we went back the way we had just come with Number 9

Journey's end, A4 No. 60009 *Union of South Africa* at the stop blocks at Swanage. In the background is estate agent Miles and Son, but this used to be the Railway Hotel and a regular haunt of volunteers in the early years. I have quaffed many a pint in there, most of it while underage! (Andrew P. M. Wright)

leading. At Redbridge we went right and headed for Romsey and Salisbury, where we went right again at Laverstock South Junction and joined the line for Basingstoke at Laverstock North Junction. Going this way turned the train and steam led all the way back to London. We rejoined the LSWR main line at Battledown Junction near Basingstoke, where we stopped in the station for water. Continuing on in the dark, we arrived back at Victoria Station at 23.00. After around thirty minutes we headed out empty stock back to Southall, routing a slightly different way around Willesden which turned the train so it would be facing the right way for Saturday's trip.

The tour to York on the Saturday was the encore but I regard the Swanage trip as my swansong and the best in terms of personal fulfilment. It was a very emotional trip as it brought my beloved loco onto my beloved railway and I was there with my family and friends, what a way to finish my time with Number 9.

Jules Hathaway

The New Lad

For as long as I can remember, Number 9 has always been present in my life. As a young boy, dad would always take me to see her whenever she ventured out and about on tours. It's ironic that when Number 9 was returned to the main line in 1973 and based at

Kirkcaldy, dad had helped out, albeit briefly. Roll forward to 15 February 2020 and here I am, as Responsible Officer, on the footplate heading south from York. An emotionally charged day as this was Number 9's last run through her birthplace, Doncaster.

As we approached with Chris Cubitt on the regulator, Fraser on the shovel, John in the fireman's seat and myself stood behind Chris, this was most definitely one of those moments when the hair on the back of your neck stood up. Thundering through on the Up fast, chime whistle echoing under the canopies, platforms 3 and 4 full, it was all over far too quickly. Goosebumps, a large smile but a lump in my throat as this was quite possibly the last main line run I would have with Number 9, and unfortunately due to the pandemic it sadly was. But what a way to see out my last trip.

For me, the highlight of being on the support crew has to be the crossing of the Forth Bridge. Many people can say they have crossed it, but not many can say they have done so on the footplate of a steam loco. There's something very special about doing this and I never get tired of those views.

You need a good sense of humour doing this. When it's not quite going your way it's always good to be able to have a joke and a laugh about it after. Like when I was let loose on the shovel for my first ever assault on Ais Gill with Number 9, and with my last round complete, the brakes came on! To my horror a lump of coal had fallen onto one of the emergency brake valves and we came to a stop right in front of the crowds that had gathered at the summit! The 'Boydie valve' was born!

In fading light on 15 February 2020, crowds line the platform as Number 9, with chime whistle blowing furiously, flies past its Doncaster birthplace for the final time. (Richard Harnetty)

Despite my relatively short time with the group, I've managed to travel to some of the most scenic parts of country. As far south as Parr over the South Devon banks and around Dawlish, to the Highlands as far as Kyle of Lochalsh. Up and over the hills to Mallaig on the line the K4 was designed for, and who could forget the most famous trip of all. A trip that almost never happened, the royal train with Her Majesty Queen Elizabeth on board for the official opening of the Borders Line to Tweedbank.

With the clock ticking after running a hot box at Warrington on our way north to work the royal train, repairs were complete just in time for us to travel to Millerhill for the big day. It was decided to send A1 *Tornado* with us 'just in case'. Thankfully it wasn't needed, and we finally arrived at 16.00 the evening before. All hands were on deck to prepare Number 9 for her big occasion. I got to my bed at 04.30 after staying up all night giving her a good wax, and, boy, were the results worth it as when daylight broke, that BR green just gleamed. Unfortunately, I didn't make the trip as tight security meant only a handful of us could travel, but I like to think I still played my part on that historic day.

The K4, in my opinion, was always living in the shadows of the A4, but what a superb little engine. A go-anywhere loco, very much at home tackling the steep climbs on the West Highland, but just as capable of taking almost double the load over the Settle and Carlisle up Ais Gill, albeit not breaking any Blue Riband records! The K4 just got its head down and away it went.

Taken from the cab of *The Great Marquess* looking forward to Number 9, the two locos cross the Forth Bridge during an engine and coach movement back to Thornton on 11 April 2013. (Chris Boyd)

Yours truly driving Number 9 up to Manuel Junction on the Bo'ness & Kinneil Railway on 7 May 2019. (Chris Boyd Collection)

It's been a whirlwind adventure, from starting out as 'The New Lad', to firing and driving both locos, the friendships, the banter and Marilyn's home cooking far from home! The memories will last for ever.

Maybe one day we will be able to see the pair working hard once again. As the saying goes, never say never.

Chris Boyd

Colin Cant

Colin was the son of a railwayman and was the last of the original Number 9 support crew, having started when the locomotive moved from Ferryhill to Thornton in 1966. He continued with Number 9 during the Lochty years and the return to the main line in 1973.

Colin worked in the finance industry all his life and in the mid-1970s his work took him away from Fife to the West of Scotland, where he would be based until 1992. During this time, he was absent from the support crew, but promptly rejoined on his return to Fife.

Colin's first trip on his return was the week-long 125th anniversary of Aberdeen station followed by a rail tour back to Fife.

Colin looks out from the support coach at Edinburgh Waverley during one of the Borders Railway opening trips. (Michael Mather)

Colin was an excellent public speaker and did the necessary for us on a number of occasions, notably at DJ's funeral, the dedication of Lindsay's seat, being interviewed by a television crew on the occasion of The Elizabethan in 1994 and, of course, the address to the haggis at our annual Burns' Supper.

Like all of us on retirement, Colin was able to commit more time to the support crew, but sadly illness caught up with him, so he had to pace himself somewhat and missed out on Number 9's last tours.

In early 2023 Colin was diagnosed with an inoperable brain tumour, the same disease that took Davy Murray from us, and he sadly passed away on 5 March at the age of seventy-one.

Always passionate about Number 9, at the end of his funeral service, at Colin's request, we left the crematorium to the sound of Number 9's chime whistle.

RIP Colin.

Michael Mather

Reminiscences of Number 9

One August day in 1991 whilst I was working in the small country Royal Bank branch in Colinsburgh the door burst open and in flew John Cameron, or 'The White Tornado' as I used to call him! 'What are you doing next Saturday?' he asked. 'I'm going to take you on my big engine – be at Edinburgh Waverley at 10am', which of course I did and so began my love of steam.

On that day there was a 'special' going out during the Edinburgh Festival, so I was in the support coach with the crew, many who have now become lifelong friends. It was a fantastic day and a wonderful experience. The crew looked after me very well giving me a guided tour of the engine and supplying me with cups of tea and the like. I went back home on the service train smelling of smoke, my hair full of coal dust and panda eyes from having my head out the window (which I know you aren't supposed to do!), all evidence that I had had a wonderful day!

It was then in June 1994 that John came back into the bank and asked if I would consider going to work for him as his bookkeeper, secretary/PA, so I thought why not as it seemed like a very interesting and varied job – which it really was for the next twenty-seven years.

At first, I didn't really have much to do with Number 9 as it was down in the Severn Valley but once it was back on the track and working again, we started to get a good number of jobs coming in. However, John was adamant that he didn't want to be bothered with all the booking procedures involved with steam trips so he insisted that he would only hire the engine out and whoever wanted it would need to organise everything else. Apparently at that time this wasn't 'how it was done' but John being John stuck to his guns and insisted that's how it was going to be going forward otherwise no engine would be forthcoming!

We then forged a very good working relationship with the Railway Touring Company and Steam Dreams. The highlight being The Great Britain tours that came up to Scotland and, of course, had Number 9 doing a few of the legs. It was always these trips that were stressful. I used to dread the days leading up to them in case something went wrong. Many a time we would get a call telling us there was a problem with a bridge or stretch of track and with John's saying of 'there's got to be a way round this, Wilma'

All smiles from John Cameron and me as we look out from Number 9's cab. (Tony Streeter, Wilma Wilson Collection)

ringing in my ears, we both had to jump into action and use his many contacts to try to resolve the issue, which to his credit he usually managed!

When the engines were up in Scotland I always used to try and go on a trip. This usually meant going over to Edinburgh to join the train and of course taking a bag of Scotch pies for the support crew. Everyone loved the pies from our local 'Barnetts' bakery so normally I had to order three dozen, which would last them for two days. Let me tell you, carrying three-dozen pies all the way to Edinburgh in a cool bag was no mean feat, but I always got a very warm welcome when I showed up at my destination!

We were also lucky enough to be up in the Highlands the summer *The Great Marques* was doing The Jacobite runs from Fort William to Mallaig, so I contacted Fraser and he arranged for us to go on a trip with them. It was a beautiful day and the scenery breathtaking, with the highlight crossing the Glenfinnan Viaduct. One to sit and recall in my old age!

Sometimes at the weekends if the engines were up in Scotland, I would go along to the yard at Thornton and give the crew a helping hand to do some cleaning. I always remember my first visit when after cleaning one side of the tender I asked if there was anything else I could do, to which someone said, 'Yes, the nameplate on Number 9 needs a shine up'. Well, when the engine is in the station the nameplate isn't too high from the platform but when it's sitting in the yard it is a fair height from the ground. I was given

Here I am with 'regulation' hi-viz boiler suit lending a hand at Thornton. (Wilma Wilson Collection)

Number 9 climbs the 1 in 70 away from Newtongrange past the Lady Victoria Pit, now home to the Scottish Mining Museum, with one of the Borders Railway opening specials. These trains ran three days a week for seven weeks. (Jules Hathaway)

instructions that I had to climb up a ladder, hook my arm round the rail and walk along the running board all whilst the engine was in steam, hissing and spitting at me, and the crew watching from the ground! Once I came down, I was given a pat on the back, and they then told me I had passed the test as not many people would do that job, so I went home that night with a spring in my step and a great sense of achievement.

The highlight for John was when Number 9 was asked to pull the royal train for the opening of the new Borders Line. I know that was a very proud day for him and the crew. I was back at Balbuthie with Mrs Cameron, but we sat and watched the coverage on TV, which was fantastic with wonderful aerial shots of the train.

It was a sad day when John decided that he wasn't going to do another overhaul on Number 9, but, as always, he never let his heart rule his head and with his vast experience and knowledge of the railways he knew the time was right for both him and the engines to retire. I can tell you I had tears streaming down my face on the days I saw the engines coming through Kilconquhar on the low loader going to their new home at the farm, but they will be well looked after and preserved for future generations. John saved Number 9 from being scrapped back in the 1960s and has given so many people the pleasure of seeing it steaming through the British countryside for many years. It is the end of an era but what an era it has been. Thanks for the all the memories, John.

Wilma Wilson

Finale

Number 9 stands at Preston station at the head of The Great Britain XII on 30 April 2019. This was the last train that Number 9 hauled to Scotland. It is double-heading with Black 5 No. 44871 and they will take the train to Edinburgh, with Number 9 taking it to Aberdeen and return the following day. (Michael Mather)

Heading its last ever train to Aberdeen, Number 9 approaches Falkland Road, between Markinch and Ladybank, with an SRPS rail tour on 5 May 2019. (Michael Mather)

For the rest of 2019 and early 2020 Number 9 worked exclusively in Englandshire, visiting many areas including Weymouth, where it is seen here, having just arrived with a Railway Touring Company rail tour from London Victoria on 22 August 2019. (Michael Mather)

An LNER Azuma heads towards York station as Number 9 stands in the sidings at the National Railway Museum on 7 March 2020, having just completed its last main line rail tour. Due to the pandemic, the loco was marooned here until July 2020 when it was moved to the East Lancs Railway to work out the remainder of its boiler ticket. (Michael Mather)

A4 Pacific No. 60009 *Union of South Africa* meets A3 Pacific No. 60103 *Flying Scotsman* at Ramsbottom on the East Lancs Railway on 30 August 2021. Number 9 was withdrawn from service a few weeks later due to a leaking boiler tube, which wasn't practical to repair as the boiler certificate was due to expire soon after. (Richard Harnetty)

On 11 May 2022 Number 9 was moved to the Bury Transport Museum, where it remained until its move to Balbuthie. While it was there, it received a full repaint. It is seen here being offloaded into the museum. (Stuart McDonald)

The Great Marquess last steamed at the North Yorkshire Moors Railway in the autumn of 2015 when it was failed with firebox problems. It was then moved to the Museum of Scottish Railways at Bo'ness, where it is seen on 4 November 2018. (Michael Mather)

On a gloomy day on the Forth and dwarfed by the structure, Number 9 makes more history by crossing the Queensferry Crossing on 22 April 2023, becoming the only steam locomotive to cross all three bridges at Queensferry. (Ryan Gray)

Above: Number 9 crosses Markinch viaduct in the 1980s. (Ian Leven). *Right*: Number 9 passes under the viaduct on its way to Balbuthie on 22 April 2023. (Ryan Gray)

With only one mile to go on its final journey to Balbuthie, and being captured on countless mobile phones, Number 9 passes through the Fife village of Kilconquhar on 22 April 2023. (Adam Clark)

After a long, eventful and record-breaking career, Number 9 rests in its new home at Balbuthie on 17 April 2024. (Marilyn Leven)

With a Gresley teak coach behind, *The Great Marquess* stands in its new home at Balbuthie on 17 April 2024. The Osprey Express headboard was made by Lindsay back in 1980 for Number 9 to carry on one of its early trips up the Highland main line. It has since been carried by both locos many times. (Marilyn Leven)

K4 2-6-0 No. 61994 *The Great Marquess* and A4 Pacific No. 60009 *Union of South Africa*, the centrepieces of John Cameron's Railway and Farming Museum at Balbuthie in Fife, on 17 April 2024. (Marilyn Leven)

Acknowledgements

Although most of the photographs in this book are from the members of the support crew and the collection of the late Davy Murray, we are indebted to the following photographers for allowing us to use photographs from their collections: Keith Verdan Anderson, Robert Avery, Ian Bowskill, Adam Clark, Peter Drummond, Ryan Gray, Richard Harnetty, Nigel Hayman, Simon Lathlane, Ian McDonald, Stuart McDonald, Edward Osmotherley, Rail Archive Stephenson, Jim Ramsay, Bill Roberton, John Turner, Nigel Valentine, John Whitehouse and Andrew P. M. Wright.

A number of photographs are credited Support Crew Collection. These have been gifted to us over the years, but unfortunately, we don't know the photographers' names. Despite repeated attempts, we have been unable to make contact with David Peaker to gain permission to use his photo of Number 9 at Bardrill Road, which was also gifted to us. These omissions will be corrected at the earliest opportunity where possible.

Final thanks to our own Marilyn Leven for proofreading, Fraser Birrell for technical information and Nick Swierklanski for our logo.